People of Anatolia

Past, Current and Future Research in the Biological Anthropology of Türkiye

Edited by

Benjamin Irvine, Yılmaz Selim Erdal & Lutgarde Vandeput

British Institute at Ankara
Monograph 58
2024

Published by
British Institute at Ankara
10 Carlton House Terrace, London SW1Y 5AH
www.biaa.ac.uk

ISBN 978-1-912090-08-2 (e-Book)
ISBN 978-1-912090-10-5 (hard cover PoD)
ISSN 0969 9007

Published with the assistance of financial contributions from
The British Institute at Ankara

Series editor: Catherine M. Draycott
Typeset by Janine Su
Index by Malcolm C. Nicholson
Distributed by Casemate UK Ltd.
Print copies by PrintForce

Contents

List of figures

List of tables

1. Introduction

Yılmaz Selim Erdal and Benjamin Irvine

Biological anthropology in Türkiye: a historical overview

The modern nation of the Republic of Türkiye is a relatively young country established to the east of Thrace, in the lands known as Anatolia. Anatolia is the ancient name for a geographical region in western Asia (also known as Asia Minor). Its full, and more proper name is the Anatolian Peninsula, and as such it is flanked by the Black Sea to the north, the Aegean Sea to the west and the Mediterranean Sea to the south. It is also at the junction of different geographical and cultural spheres such as the Levant, Mesopotamia, the Zagros, the Caucasus and Europe. Although the earliest evidence for human occupation in Anatolia dates back to the Palaeolithic period, there is almost no information on the Palaeolithic people. Our knowledge on this earliest phase of human occupation in Türkiye is mainly based on material culture such as stone artefacts and some rock art, as well as the one of the literal impacts of human existence – footprints. However, both cultural and biological data on the history of human occupation comes mostly from Late Epipalaeolithic and the Neolithic period, an era marked by the beginning of early sedentary life and the domestication of plants and animals. As part of the core region of the Neolithic transition, which is characterised by the earliest permanent settlements and the domestication of some plants and animals, Anatolia and its surrounding areas played an important role in shaping human populations in southwest Asia and then later in Europe. This means that Anatolia is an important region for archaeological and anthropological studies to understand the emergence and development of different cultures and civilisations.

Anthropological studies and/or the interpretation of skeletal remains of Anatolian people started in the latter stages of the Ottoman Empire in the last quarter of the 19th century. These studies did not aim to interpret the osteobiographies of past people, rather their aim was understanding their origin, and more generally identifying their 'race' in parallel with the common scientific approach of this period. In other words, the first biological anthropological studies on Turkish skeletal material were a product of the times, where bioarchaeological and osteobiographical aspects were not dealt with or engaged with, and skeletons (particularly crania) were simply measured in an attempt to classify them. Therefore, it can be stated that Turkish skeletal materials were initially subjected to techniques, methods and anthropological practices imported from abroad. One of the first studies, for example, was by the Austrian physician Augustin Weisbach, who published an article in 1873, 'Die Shadelform der Türken' in the *Mitteilungsblatt der Anthropologischen Gesellschaft*. In it he explained, following the measurement of 78 crania, that the Turkish 'race' has a brachycephalic cranium (see also Üstündağ, Yazıcıoğlu 2014). Studies by scientists such as Rudolf Virchow, Felix von Luschan and E. Fisher were based on identifying and classifying the 'races' of people living in Anatolia, in modern Türkiye, and the origin of the Indo-Europeans. These pioneering types of research influenced later studies until the 1990s.

Turkish anthropologists did not discuss the concept of race from a theoretical viewpoint. Later, an understanding of the plasticity, adaptability and morphological instability of the human body and its skeleton, and especially the crania, resulted in the logical rejection of the 'race' concept. Turkish scientists slowly abandoned the use of 'race' as a concept. Some anthropologists, though, did continue to analyse human remains as part of the typology of humans and human variation, and crania were mainly used for these types of studies. While Lawrence J. Angel's studies in the 1950s used the concept of 'cranial type' rather than 'race', he explained that

> the concept of morphological types is real in any given investigation but only extends to (a), that these types generalize the phenotypic effect of genetic linkage; (b) that allowance can be made for the effects of crossing over and incorporation of new genetic possibilities into the types originally defined, and (c) that the environment remains stable (Angel 1951a: 21).

He attempted to infer genetic origin by utilising data from the skeletal remains. In studies after the 1940s, 'race' became nothing more than an attempt to make sense of biological diversity rather than the origin of the Anatolian people. Angel (1951b: 478) stated that 'there are two ways of looking at ancient skeletal material: merely as a static record of craniological and osteological characteristics of an early people; or as a dynamic interpretation of biogenetic potentials in a population sample'.

Whilst we want to acknowledge that the concept of 'race' is an important issue to consider in biological anthropology and its research history, it is not the focus of this volume. We are not able in this volume to do justice to such a complex topic, and therefore find it inappropriate to engage with it in depth. Indeed, beyond this historical overview of Turkish anthropology's brief flirtation with identifying and classifying 'race' (a product of the broader anthropological thought of its time) the concept of 'race' has had little influence on biological anthropological research in the country. This is especially true relative to regions like Europe and in particular North America. Whilst the techniques and methodologies to measure and classify human skeletal remains were imported to Türkiye, the racist ideologies behind them (as in North American and European scholarship) were not – or perhaps more accurately they were not implemented and applied in the same way, and had little to no impact on modern Turkish society. The concept of 'race' from cranial measurements was rather quickly abandoned in Turkish anthropology, likely due to a lack of any ideological need to pursue bioarchaeological research questions. This would have also been due to the influence of anthropologists like Lawrence J. Angel, Wilton M. Krogman and Muzaffer Süleyman Şenyürek, who were mainly focused on the bioarchaeological aspects of the Anatolian people during their research from the 1950s to the 1970s. They helped to shape the focus and direction of biological anthropology in Türkiye. Lawrence J. Angel, in particular, mainly focused on growth, stature, hereditary and infectious disease, skeletal trauma and palaeodemographical patterns in the Eastern Mediterranean region in relation to (the emergence of) agriculture and the development of civilization. Angel's pioneering studies on anaemia, which hypothesised a connection with malaria wherein it increased in rate in relation to the development of agriculture, were some of the first studies examining human-environment interaction. Angel's (1996) paper on porotic hyperostosis was mainly focused on balanced polymorphism between hereditary anaemias and malaria in the Mediterranean area, and Türkiye was one of the main regions of this study. He also proposed a connection between agriculture, anaemia, stature and anthropological demographic characteristics (Angel 1984). Some of his most famous works come from the analysis of human remains from the sites of Çatalhöyük and Karataş (Angel 1971). Together with Angel, studies on human remains were also conducted by Muzaffer Süleyman Şenyürek. Şenyürek made significant contributions to Turkish biological anthropology studies in the 1940s and 1950s in different fields of investigation such as trepanation, intentional headshaping and dental morphology. These studies accelerated the bioarchaeological approach until the 1970s. However, there was a period of discontinuation in biological anthropology studies, especially in the field of bioarchaeology, following Şenyürek's tragic death in a plane crash in 1961 and Angel's departure from Türkiye.

Whilst there have been numerous studies on the bioarchaeology of human populations – and this has been the primary approach since around the middle/latter third of the 20th century – biological anthropology studies in Türkiye have not completely abandoned the study of cranial typology and the origin of ethnic groups. However, rather than simply classifying 'race', these kinds of examinations of cranial types have been conducted to understand the demographic structure and the mobility of populations and examining genetic and phenotypic drift and admixture – especially in the days before ancient DNA (aDNA) analysis. For example, by investigating the Karataş human remains, it was suggested that speakers of Indo-European languages moved westward and reached Europe (Angel 1970: 254). In the same period, Cappieri (1970) explored the population history of Anatolian people by analysing cranial phenotype. These studies suggested that cranial morphology could reflect genetic relatedness and gene flow, and, thereby, the population dynamics of Anatolia. They explained the change of population phenotypes either as the result of evolutionary dynamics (Cappieri 1970; 1972) or human movement and migration (Angel 1970). Although it is only possible to obtain relatively abstract information about population structure using cranial measurements, anthropologists observed an increased genetic diversity and mobility (through observations of phenotypes) in the Early Bronze Age period (ca 3200–2000 BC) of Anatolia. A study by Saatçioğlu (1982: 193–209), titled 'A Survey on the Racial Types of Anatolian Skeletal Remains', investigated human diversity in Anatolia and changes in the population structure from the Neolithic period onwards. Cranial typology was also further investigated using analysis of blood type prevalence (Saatçioğlu 1978). As part of investigating 'racial characteristics' (not in the sense of 'race'), anthropologists frequently referred to cranial types in the 1980s–1990s in Türkiye. Biological anthropologists used cranial types and their subsequently identified 'racial characteristics' to understand the phenetic history of Anatolia. This data suggested that the Mediterranean type was common in the Neolithic, after which the frequency of the Alpine type increased, and finally the Dinaric type was incorporated into the human phenotypology of Anatolia. For example, following analysis of the human remains, Angel (1970) suggested that the Karataş people diverged (genetically) from their likely local ancestors towards the probable direction of the Macedonian Neolithic group and those of the later Aegean populations. He also proposed a genetic relation with the east and possibly also with Cyprus. Cappieri (1969; 1970; 1972) tried to understand whether the Hittites were local or immigrant populations of Anatolia by examining the craniometrics from skeletal remains discovered in the Hittite heartlands and their surrounding regions, as well as examining the craniometrics of pre-Hittite, Hittite and post-Hittite skeletal remains. Rather than determining the 'races' of Anatolian societies, Cappieri recognised the presence of human population types such as the Proto-Mediterranean Type and the Anatolian Regional Type by using craniometry and examining the morphology of the skulls. He considers the change of Anatolian human societies on an evolutionary basis to be a result of their development and progression from the Palaeolithic period.

After the 1990s, cranial types and morphology were no longer used to classify populations with regard to 'race' or even 'racial characteristics'; instead, biological anthropology studies in Türkiye mainly focused on the reconstruction of population history (Wittwer-Backofen 1985; Eroğlu 2005; Eroğlu, Erdal 2006; 2009), burial customs, daily activities, human health and reconstructing the dietary habits of ancient Anatolian populations. These 'modern' studies mainly focused on description and interpretation by applying contemporary knowledge and theoretical aspects. However, although they are currently few in number, there have also been studies coming out of Türkiye proposing new methodologies.

Current research and this edited volume

This edited volume brings together some of the current research foci and trends of biological anthropology in Türkiye, and was inspired by the workshop 'Physical Anthropology in Anatolia', which was held at the British Institute at Ankara (BIAA) on the 8th of November 2019. One of the main aims of the workshop was to present the current state of affairs of biological anthropological studies in Türkiye, and to provide a small section of the studies in the field as primary examples. Studies containing the general evaluations of Turkish researchers who are specialised in their fields were invited, taking into account the different fields of biological anthropology. Therefore, to some extent, the articles in this book will also present the current state of biological anthropological research in Türkiye.

Papers within this edited volume cover a multitude of topics which help to demonstrate the strength and variety of research currently being performed in Türkiye by both domestic and foreign researchers. Included are papers on some of the 'basics' of biological anthropology, such as estimating sex and stature and identifying pathologies, and placing them within their past contexts through case studies. Also included are more detailed studies and large, diachronic datasets to investigate patterns through time and space. It is important to make this kind of data and analysis available to the wider archaeological community, as well as Turkish students of archaeology, especially the subdisciplines of biological anthropology and bioarchaeology.

The papers by İsmail Özer and Mehmet Sağır deal with some of the fundamentals of biological anthropology, namely establishing two of the key demographic characteristics of estimating sex and stature, respectively. Handan Üstündağ highlights the importance of considering the 'osteological paradox' in biological anthropological research, using porotic hyperostosis as a case study. Serpil Eroğlu provides an overview of biological distance studies performed over the years on skulls from Anatolian populations, and how this subdiscipline is integrating with modern techniques like aDNA studies. Staying with the skull, Valentina D'Amico investigates intentional headshaping from prehistory to the present day in Anatolia, providing an overview of previous research, and also considering ethnic and genetic factors that can influence this phenomenon. Aydoğan et al.'s paper on aDNA in Anatolia helps to provide information about the application of scientific approaches to biological anthropology, and how these can be used in collaboration with the more traditional techniques used by Eroğlu, Özer and D'Amico. The final three papers deal with reconstructing dietary habits and subsistence strategies, and with examining diachronic changes in them. Yılmaz Selim Erdal's paper draws upon previous studies and a re-evaluation of the currently known data of activity-related dental and skeletal pathologies to provide a comprehensive investigation into the effects that Neolithisation and the adoption of a fully agricultural way of life had on human societies and their skeletons. Melis Koruyucu's paper complements this by taking a diachronic look into caries rates and frequencies from the Neolithic period to the present day, providing insights into what the data can tell us about dietary habits, subsistence changes and food preparation techniques. Benjamin Irvine's paper wraps up the volume by applying a scientific approach, through stable isotope analyses, to diachronically investigate dietary habits and changing subsistence models in Anatolia, from the Neolithic to the Late Byzantine period.

Whilst the individual papers in this edited volume are, in themselves, not all necessarily inter-/multidisciplinary, reading them all produces a holistic overview of biological anthropological research in Türkiye. Several of them also examine large-scale diachronic changes and help to highlight the importance of such holistic approaches and methodological considerations; together, this collection provides a 'big picture' look at trends in the modern research combining more traditional macroscopic biological anthropological analyses with more modern techniques such as biomolecular, biogeochemical and genetic analyses.

Bibliography

Angel, J.L. 1951a: *Troy, The Human Remains*. Princeton, Princeton University Press

— 1951b: 'Troy, the human remains' *American Journal of Physical Anthropology* Supplemental Monograph 1.40: 478–97

— 1966: 'Porotic hyperostosis, anemias, malarias, and marshes in the prehistoric Eastern Mediterranean' *Science* 153: 760–63

— 1968: 'Human remains at Karatas' *American Journal of Archaeology* 72.3: 260–62

— 1970: 'Human skeletal remains at Karatas' *American Journal of Archaeology* 74.3: 253–59

— 1971: 'Early Neolithic skeletons from Catal Hüyük: demography and pathology' *Anatolian Studies* 21: 77–98

— 1984: 'Health as a crucial factor in the changes from hunting to developed farming in the Eastern Mediterranean' in M.N. Cohem, G.J. Armelagos (eds) *Paleopathology at the Origin of Agriculture*. Orlando, Academic Press: 51–73

Cappieri, M. 1969: *The Mediterranean Race in Asia before the Iron Age*. Miami: Field Research Projects Coconut Grove

— 1970: 'The Anatolian of the Late Neolithic and Chalcolithic Age' *Belleten* 34: 509–55

— 1972: *Genotypical and Phenotypical Homogeneity of the Ancient Population of Anatolia*. Miami: Field Research Projects Coconut Grove

Eroğlu, S. 2005: *Anadolu'da Bazı Eski İnsan Topluluklarında Biyolojik Uzaklıkların Belirlenmesi*. PhD thesis, Hacettepe Üniversitesi, Ankara

Eroğlu, S., Erdal, Y.S. 2006: 'İkiztepe insanlarının Anadolu ve Anadolu'ya yakın topluluklarla biyolojik ilişkisi' *Arkeometri Sonuçları Toplantısı* 21: 41–50

— 2009: 'Diş ve kafatası morfolojisine dayanarak üç eski Anadolu topluluğunda biyolojik uzaklıkların belirlenmesi [The determination of biological distance based on morphological traits of the human teeth and skulls in three ancient Anatolian populations]' *Hacettepe Diş Hekimliği Fakültesi Dergisi* 33.3: 78–90

Saatçioğlu, A. 1978: *AB0 Genleri Yönünden Türkiye'nin Yeri ve bu Ülkedeki Gensel Çeşitlilik Üzerine biyometrik bir inceleme*. Ankara, Ankara Üniversitesi Yayınları

— 1982: 'A survey on the racial types of Anatolian skeletal remains' Ankara Üniversitesi DTCF Dergisi 30: 193–209

Üstündağ, H., Yazıcıoğlu, G.B. 2014: 'The history of physical anthropology in Turkey' in B. O'Donnabhain, M.C. Lozada Cerna (eds) *Archaeological Human Remains*. Cham, Springer: 199–211

Wittwer-Backofen, U. 1985: 'Anthropologische untersuchungen der Nekropole Ikiztepe/Samsun' *Araştırma Sonuçları Toplantısı* 3: 421–28

2. Sex estimation and gender in ancient Anatolian populations

İsmail Özer

Introduction

In living people, sex is a discrete trait determined by an individual's genetic makeup. There are easily identifiable characteristics that can be used to categorise anyone as male or female. Unfortunately, identification is much more difficult in the human skeleton because the variety in shape and size of all features form a continuum. This challenge in estimating sex is exacerbated by the fact that the remains are often fragmentary; there might also be differences in the expression of specific traits at the population level and in the identification of certain traits depending on the examiner's experience (İşcan, Steyn 2013).

Precision is the degree to which a determination matches reality. For example, to age a mandible as a subadult might be considered *accurate* while also being *imprecise*. How accurately and precisely can an anthropologist determine the sex, age, size and lineage of human skeletal remains? There is no easy answer to the question. Any identification of a biological parameter such as sex, age, stature or ancestry is, in fact, a matter of probability. The likelihood of a particular identification being correct depends on many different factors which generally need to be taken into account before proceeding with the analytical methods themselves (White et al. 2012).

The most effective sexual indicators of the skeleton only begin to develop during adolescence, and many are not fully expressed until adulthood. Sexing of immature skeletal remains is, therefore, extremely difficult, and the results are preliminary at best. Recent research has also shown that skeletal changes after adulthood can mask dimorphism. For example, in his 2005 article, Walker reported an association between age and the larger shape of the sciatic notch. The notch tends to be wider (more feminine) in younger men and women, but as people get older it tends to become more masculine (i.e. narrower). Vitamin D deficiency has been cited as a possible cause of this phenomenon. Vance et al. (2011) found that many postcranial measurements in white women and men increased significantly with age, while black women showed no change. In that study population, the white people tended to be more prone to osteoporosis. Height may be affected by normal degenerative changes, microfractures on joint surfaces, and changes in the relationship between cortical and endosteal bone. Thus, sexual dimorphism across a lifespan may not be as stable as we think, which could affect our ability to determine an individual's sex (İşcan, Steyn 2013).

Palaeodemographic analysis of skeletons

Palaeodemography in its modern form is only a few decades old. Certainly, bones and teeth have long been used to estimate the sex and age-at-death of skeletons at crime scenes and archaeological sites. Much of this work, however, is not palaeodemography if one refers to the rigorous study of the demographic structure of previous populations. Reports of excavations from the early to mid-20th century often include data indicating the age-at-death and sex of the individuals interred in the graves, as well as the positions of the body and associated artefacts. Accurate estimates of the age-at-death and sex of human skeletal remains are the basis of studies into past adaptations and demographic histories. Age and sex differences in diet, disease, activity patterns and burial practices are all accessible through studies of human skeletal materials. Skeletal studies can also offer diachronic perspectives on human demography that are not available from any other source. The accuracy of the age-at-death assigned to the skeletons was of great concern, as it is today, and some parts of the skeleton are more effective than others for estimation purposes. Sex estimation is also a priority for the further analysis of ancient unidentified human remains, as all identification techniques for males and females are markedly different (Buikstra, Ubelaker 1994; Adams 2007; White et al. 2012; Milner et al. 2019).

Sex estimation

Estimation of sex based on human skeletons is vital in forensic medicine, as well as in anthropology and archaeology. For this reason, many methods have been developed to determine the sex of skeletons. All of these applied techniques are very different for estimating male or female sex. The determination of sexual dimorphism

Term	*Symbol*	*Should be read as*	*Meaning*
Female	♀	Female	The researcher has full confidence in determining the sex of the remains
Male	♂	Male	
(Female)	(♀)	Probably female	The researcher is not fully confident in the determination but feels the remains are probably the stated sex
(Male)	(♂)	Probably male	
Female ?	♀ ?	Possibly female	The researcher has no confidence in the determination but believes the available evidence suggests the indicated sex
Male ?	♂ ?	Possibly male	
Indeterminate		Sex indeterminate	The remains were analysed, but the diagnostic morphology is insufficient to allow the determination of sex
Unknown		Unknown sex	

Tab. 2.1. Terminology and symbols used to determine sex from the human skeleton (after White et al. 2012).

from osteological material is usually related to differences in body volume and muscle structure, as well as the ability of women to give birth.

How exact can a forensic scientist be when estimating sex? Krogman (1939) sexed a sample of 750 adult skeletons (white and black, male and female) from the Hamann-Todd collection. Success rates were as follows: when the entire skeleton was present (100%), pelvis alone (95%), skull alone (92%), pelvis plus skull (98%), long bones alone (80%), long bone plus pelvis (98%). Stewart (1951) estimated that for the entire skeleton or pelvis, he could be right 90–95% of the time; for single adult skulls, this reduced to about 80%. In his previous work, Stewart (1948) also mentioned that Hrdlička was 80% accurate with a single skull, but when the mandibles were present, he achieved 90% accuracy. In a series of 100 adult black-American skeletons that were sexually identified following a full skeletal inspection, Stewart got a score of 94% accuracy, but in that same series using the skull and mandible, he was right only 77% of the time. Duric et al. (2005) reported 100% accuracy of skeletons from a forensic environment in the Balkans using pelvic and cranial features, although their sample only included males, which may have biased the results. From these studies, it therefore can be deduced that it is not possible to make the correct estimate in all cases, even with a complete adult skeleton (İşcan, Steyn 2013). Due to the uncertainties associated with determining sex from human skeletal remains, knowing and understanding some of the terms that express both sex determination and the researcher's confidence in the determination would be useful (Tab. 2.1) (White et al. 2012).

There are two methodological approaches to estimating the sex of skeletal remains: morphological and metric. Most of the older studies which examine sex differences of the skeleton (primarily of the skull and pelvis) have focused on describing morphological characteristics. These descriptions focus on the shape and bone configurations that are visible macroscopically. The advantages of this approach are many, especially when a particular shape is consistently observed despite different populations and chronologies. However, there is some subjectivity, and it is difficult to know exactly how accurate the results will be. Many of the previously held assumptions about morphological differences are currently being questioned, and their existence and accuracy in sexual distinctiveness are being reassessed using modern morphometric techniques (İşcan, Steyn 2013).

Sex estimation using the morphological method

In all human groups, changes in growth during adolescence cause marked differences between male and female os coxae. A woman's pelvic outlet is enlarged in proportion to her size, while the male growth/growth pathways generally follow a pre-pubertal pattern (Coleman 1969). Therefore, adolescent hip bones with uniquely female characteristics can be safely diagnosed as female. However, a male model observed in a juvenile os coxa should not be considered conclusive, as the remains may represent a female whose pelvic dimensions have not yet taken on adult proportions. Conversely, male skulls can maintain a slim female appearance during puberty. The skulls of adolescents with masculine features can therefore be diagnosed as masculine with greater certainty than a more slender skull. The use of skull features is further complicated by age-related changes in dimorphism into adulthood. In both men and women, the skull tends to have an increasingly 'masculine' morphology with age (White et al. 2012). To date, there are no standards for diagnosing adolescent sex that osteologists have reached a consensus on as acceptable (Saunders 1992).

Determining sex based on parts of the skull follows the observation that men tend to be taller and more robust than women. Male skulls are generally more robust than female skulls (see e.g. White et al. 2012: 409, fig. 18.20), with more pronounced supraorbital ridges, a more prominent glabellar region, and heavier temporal and cervical lines. Male foreheads and parietals tend to be less directed than those of females. Males tend to have relatively large and wide palates, more square eye sockets, larger mastoid processes, larger sinuses and larger occipital condyles than females. Compared to female mandibles, male mandibles are characterised by a more square jaw, a more gonial angle, deeper lower jaw branches and rougher muscle attachments. In all cases, a five-point scale is used, which is interpreted as follows: 1 = hyper-feminine, 2 = female; 3 = undefined; 4 = male; 5 = hyper-masculine. This scale helps the researcher to standardise observations for comparison purposes (see White et al. 2012: 410, fig. 18.21).

The pelvis is the most dimorphic bone in the human skeleton, and has been studied extensively for the purpose of estimating sex on unknown skeletal remains. Traditional methods for determining the sex of the pelvis or its parts are based on the following trends: the sacrum and ossa coxae of males are larger and more robust than those of females; the pelvic entrances of females are relatively wider than those of males; the sciatic notches of female ossa coxae are relatively larger than those of male bones; females have relatively longer pubic parts of the coxae, including the upper pubic branch, than males; the subpubic angle formed between the lower edges of the two lower pubic branches is greater in females than in males; the preauricular sulcus is more common in

Researcher	*Bone / Region*	*Researcher*	*Bone / Region*
Hanihara 1958	Long bones	İşcan, Ding 1995	Femur
Hanihara 1959	Cranium and scapula	Trancho et al. 1997	Femur
Singh et al. 1975	Tibia	Steyn, İşcan 1997	Femur and tibia
Black 1978	Femur	King et al. 1998	Femur
Tanaka et al. 1979	Cranium	İşcan et al. 1998	Humerus
DiBennardo, Taylor 1979	Femur	Özer 1999	Whole skeleton
Hanihara 1981	Tooth	Mall et al. 2000	Femur
DiBennardo, Taylor 1982	Femur	Günay and Altınkök 2000	Foramen magnum
Taylor, DiBennardo 1982	Femur	Mall et al. 2000	Arm bones
DiBennardo, Taylor 1983	Post cranium	Asala 2001	Femur
Schulter-Ellis et al. 1983	Pelvic	Purkait 2001	Ulna
İşcan, Miller-Shaivitz 1984	Femur	Purkait, Chandra 2002	Femur
İşcan, Miller-Shaivitz 1984	Tibia	Özer, Sağır 2002	Cranium
Schulter-Ellis et al. 1985	Pelvic	Özer, Sağır 2003	Long bones
MacLaughlin, Bruce 1985	Femur	Purkait 2003	Femur
İşcan, Miller-Shaivitz 1986	Femur and tibia	Purkait, Chandra 2004	Femur
Dittrick, Suchey 1986	Femur and humerus	Sakaue 2004	Long bones
Nakahashi, Nagai 1986	Post cranium	Özer, Sağır 2004	Cranium
Ren 1987	Scapula	Özer, Katayama 2006	Scapula
Yang et al. 1988	Mandibula	Özer et al. 2006	Femur
Zheng, Pang 1988	Tibia	Özer, Katayama 2008	Femur
Liu 1989	Femur	Soni et al. 2010	Femur
Liu et al. 1989	Lower extremities	De Sousa et al. 2010	Femur
Liu 1989	Upper extremities	Pandya et al. 2011	Femur
Li 1991	Cranium	Chowdhury et al. 2012	Femur
Holland 1991	Tibia	Bhosale, Zambare 2013	Femur
Holman, Benneth 1991	Arm bones	Özer 2014	Arm bones
Song et al. 1992	Cranium	Özer et al. 2014	Tibia
Inoue et al. 1992	Cranium	Divakar et al. 2016	Cranium
İşcan et al. 1994	Tibia	Usta 2019	Metacarpal and metatarsal
İşcan et al. 1995	Cranium		

Tab. 2.2. Selected studies determining sex utilising discriminant function analysis.

females than in males. One consequence is that the atrial area of the female ilium is higher than that of the male ilium, although the sexual dimorphism of the atrial surface itself is insufficient for accurate sex determination. The acetabulum also tends to be relatively larger in males (see White et al. 2012: 417, fig. 18.26). Sometimes, not all traits may be highlighted equally – one trait may be more masculine in a particular pelvis while another trait may be more feminine. The front part of the pelvis is more closely associated with childbirth, while the rear part is more closely linked to the differences in the mode of locomotion between males and females (Bruzek 2002).

Sex estimation using the metric method

Some of the most powerful methods for determining sex from the skeleton rely on the application of statistical analysis to osteological material. Due to the differences between populations, there is a need to develop specific standard methods for estimating each sex. The analysis of discriminant functions as a mathematical approach allows the development of sexing methods using all measurements taken from the skeleton. Determining sex by analysing discriminant function provides information about each bone, which is very useful in post-disaster and forensic cases. Methods for determining sex by discriminant skeletal analysis have been described in several populations by numerous authors (see Tab. 2.2).

Özer (2014) produced one of the pioneering comparative studies on sex estimation using the discriminant function in Türkiye, examining ancient Japanese and Anatolian populations. In this study, a sample of arm bones (humerus, radius and ulna) was examined from 151 adult skeletons of the Yoshigo population (Aichi, Japan) and 156 adult skeletons from Dilkaya (Van, Türkiye). Following the analysis of a total of 15 arm bones, all measurements showed statistically significant differences between the sexes. The epicondylar width of the humerus was shown to be the most reliable measurement for estimating sex in the study populations, with the general sex reliability range from this measurement being 70.5–86.3% for the Yoshigo sample population and 70–92.6% for the Dilkaya sample population (Özer 2014).

Many researchers claim that metric methods are superior to morphological methods because they are objective, and result in lower intra- and interobserver rates, regardless of an individual's experience (İşcan, Steyn 2013; Koca Özer et al. 2014). Cranial and postcranial metrics are also more easily subjected to statistical analysis, facilitating comparisons within and between populations, as well as multivariate, probabilistic and error analysis. On the other hand, metric analyses usually require intact skeletal elements and are based on size, which is known to vary widely between populations. Beyond this, however, the preferred methods for determining sex vary depending on the anthropologist's subjectivity and experience, as well as the skeletal elements available for analysis. Some variables are effective as a discriminator, such as the width of the elbow or knee joints. Scoring the skull, both metrically and morphologically, provides correct estimates 80–90% of the time. The dimensions of long bones can be as good as the skull, and are around 90% accurate when population-specific formulas are used for some of the larger bones, such as the femur and humerus (İşcan, Steyn 2013). Sections of the diaphyseal parts of the bones, which should also reflect mechanical stress, were found to not be as good a discriminator as the epiphyseal parts, and the distal part of the humerus or femur was found to be better than the proximal part. In mixed graves wherein only well-preserved long bones remain, the humerus or tibia may be used for better sex determination than when using the ulna or femur. The differences between arms and legs can result from the different mechanical constraints related to habitual workloads and physical activities. In agricultural people, mechanical stress mainly affects the upper body, and as a result, the circumference of the shaft of the arms is more pronounced in men than in women. In the leg, only a few dimensions are linked to muscle use and movement that exhibits the same behaviour (Tab. 2.3).

Gender in ancient Anatolian populations

The terms 'sex' and 'gender' are increasingly confused in the anthropological and medical literature. They do not refer to the same thing, they are not synonyms and they should not be used interchangeably. Gender is an aspect of a person's social identity, while sex is related to a person's biological identity. Preserving this distinction is important for biological anthropologists and in human osteology (Walker, Cook 1998). In an archaeological context, it is often possible to determine sex by analysing skeletal remains and gender roles through studies of material culture and context. Anthropologists examine the relative positions in a society of men, women and children by identifying and examining the differences in power and authority that are manifested in the material and skeletal remains. These differences can survive in the physical record, although they are not always immediately apparent and often need to be interpreted. The relationship between the sexes can also affect the relationships between other social groups such as families, different socio-economic and socio-political classes, age groups and religions.

Region	*Discriminant function formulae*	*Accuracy*	*Source*
Cranium	(0.50255 x bizygomatic breadth) + (-0.07786 x basion-nasion length) + (0.24989 x mastoid height) + (0.19553 x nasal height) + (0.24263 x basion-bregma height) + (-0.15875 x minimum frontal breadth) + (-0.13224 x biauricular breadth) + (0.21776 x glabella-occipital length) + (-0.09443 x frontal chord) + (-0.08327 x parietal chord) + (-0.13411 x occipital chord) – 81.1812 Sectioning Point: 0	90.01% (tested on modern North American white adults)	Spradley, Jantz 2011
Mandible	(0.15798 x maximum ramus height) + (0.21951 x bigonial width) + (0.06335 x mandibular length) – 35.0107 Sectioning Point: 0	80.80% (tested on modern North American white adults)	Spradley, Jantz 2011
Tooth	(2.572 x U1 mesiobuccal-distolingual) – (1.281 x U1 disto-buccal-mesiolingual) – (1.207 x L2 mesiobuccal-distolingual) + (1.695 x LC mesiobuccal-distolingual) – 16.136 Sectioning Point: 0	83.30% (tested on modern Turkish adults)	Karaman 2006
Scapula	(0.024 x maximum scapular height) + (0.043 x maximum scapular breadth) + (0.316 x glenoid cavity height) + (0.145 x glenoid cavity breadth) – 22.838 Sectioning Point: 0	95.00% (tested on Anatolian medieval adults)	Özer et al. 2006
Humerus	(0.276 x epicondylar breadth) – 16.282 Sectioning Point: -0.1915	87.40% (tested on Anatolian medieval adults)	Özer 2014
Radius	(0.660 x caput radii thickness) – 13.893 Sectioning Point: -0.0425	92.60% (tested on Anatolian medieval adults)	Özer 2014
Ulna	(0.103 x maximum ulna length) – 24.219 Sectioning Point: 0.032	82.8% (tested on Japanese Late Jomon adults)	Özer 2014
Femur	(0.024 x maximum femur length) + (0.107 x circumference) + (0.160 x condyle breadth) – 30.413 Sectioning Point: -0.1745	100% (tested on Japanese Late Jomon adults)	Özer, Katayama 2008
Patella	(0.292 x maximum height) + (0.080 x maximum height of articulating facet) – 14.342 Sectioning Point: 0	84.2% (tested on South African white adults)	Bidmos et al. 2005
Tibia	(0.007 x maximum tibia length) + (0.077 x transverse diameter at the nutrient foramen) + (0.099 x (sagittal diameter at the nutrient foramen) + (0.121 x minimum circumference) – 16.192 Sectioning Point: 0.79	90.20% (tested on Anatolian medieval adults)	Koca Özer et al. 2014
Calcaneus	(0.352 x dorsal articular facet breadth) + (0.228 x middle breadth) (0.063 x minimum breadth) – 18.133 Sectioning Point: 0.065	89.10% (tested on South African white adults)	Bidmos, Asala 2003

Tab. 2.3. Sex determination utilising discriminant function analysis for selected parts of the adult human skeleton.

	Participant							
	According to women (%)				According to men (%)			
Domestic works behaviours	*Females*	*Males*	*Suitable for both*	*Not suitable for both*	*Females*	*Males*	*Suitable for both*	*Not suitable for both*
Cooking	76	10	8	6	92	2	4	2
Washing dishes	83	5	7	5	89	2	7	2
Cleaning the house	76	18	4	2	91	3	5	1
Dressing the house	79	10	11	0	88	5	7	0
Householder	12	64	22	2	4	82	14	0
Cycling	11	33	54	2	8	35	56	1
Using weapons	6	79	3	12	3	86	9	2
Holidaying alone	3	75	7	15	2	84	4	10
Authoritarian	12	57	20	11	6	62	29	3
Leader	9	49	40	2	4	56	37	3

Tab. 2.4. Results of a perception survey about gender roles in modern Turkish society (following Vatandaş 2007).

Vatandaş (2007) highlighted that environmental conditions and biological differences between the sexes represent the divisions of labour in the community, explaining that this situation arose in the past and has continued into the present. In this traditional structure, women take on roles in companies and professional groups that do not require physical strength, while men move towards social life, work and professional fields that require hard work and development and, in some instances, physical violence. Some conclusions in Table 2.4 have been drawn from Vatandaş's (2007) study of the more recent/modern Turkish population, which examined this problem. Data was collected using a questionnaire, which was completed by 3454 people from a total of 197 settlements – 41 urban and 156 rural – which were deliberately selected to be representative of the population at large. In the table, we see that men and women also assume the social roles assigned to them (Vatandaş 2007).

In palaeoanthropological studies of ancient Anatolian populations, assessments have been made about gender roles in society based on sexual differences observed in skeletal populations. For example, following palaeopathological analysis of the İkiztepe skeletons, Erdal (2010) revealed the presence of fatal skull injuries in males, and none in females. It was interpreted that the reason for this difference was males being involved in interpersonal violence, possibly including organised violence such as warfare. A separate study by Erdal (2008) presents grooves on the anterior teeth of five female skeletons found at Kovuklukaya, which was associated with weaving activities. Furthermore, based on the presence of medial epicondylitis seen in the humerus of people from Kovuklukaya, it was argued that carpet weaving was a very important part of daily life at the settlement (Erdal 2008).

As in the examples briefly mentioned above, the roles assigned to individuals by society shape people's daily lives biologically and culturally. In palaeoanthropological studies, we try to understand gender roles that affect biological and cultural life through interpretation and analysis of human skeletal material.

Bibliography

Adams, B.J. 2007: *Forensic Anthropology*. New York, Chelsea House Publishers

Asala, S.A. 2001: 'Sex determination from the head of the femur of South African whites and blacks' *Forensic Science International* 117: 15–22

Bhosale, R.S., Zambare, B.R. 2013: 'Sex determination from femur using length of femur in Maharashtra' *Journal of Dental and Medical Sciences* 3.4: 1–3

Bidmos, M.A., Asala, S.A. 2003: 'Discriminant function sexing of the calcaneus of South African whites' *Journal of Forensic Science* 48.6: 1–6

Bidmos, M.A., Steinberg, N., Kuykendall, K.L. 2005: 'Patella measurements of South African whites as sex assessors' *HOMO – Journal of Comparative Human Biology* 56: 69–74

Black, T.K. 1978: 'A new method for assessing sex of fragmentary skeletal remains: femoral shaft circumference' *American Journal of Physical Anthropology* 48: 227–32

Buikstra, J.E., Ubelaker, D.H. 1994: *Standards for Data Collection from Human Skeletal Remains*, 3rd Edition. Fayetteville, Arkansas Archaeological Survey Research Series No. 44

Bruzek, J. 2002: 'A method for visual determination of sex, using the human hip bone' *American Journal of Physical Anthropology* 117: 157–68

Chowdhury, M.S., Naushaba, H., Mahbubul Mawla Chowdhury, A.H.M., Khan, L.F., Ara, J.G. 2012: 'Morphometric study of fully ossified head and neck diameter of the human left femur' *Journal of Dhaka National Medical College* 18.2:9–13

Coleman, W.H. 1969: 'Sex differences in the growth of human bony pelvis' *American Journal of Physical Anthropology* 31: 125–51

De Sousa, E.B., Fernandes, R.M.P., Mathias, M.B., Rodrigues, M.R., Ambram, A.J., Babinski, M.A. 2010: 'Morphometric study of the proximal femur extremity in Brazilians' *International Journal of Morphology* 28.3: 835–40

DiBennardo, R., Taylor, J.V. 1979: 'Sex assessment of the femur: a test of a new method' *American Journal of Physical Anthropology* 50: 635–37

— 1982: 'Classification and misclassification in sexing the black femur by discriminant function analysis' *American Journal of Physical Anthropology* 58: 145–51

— 1983: 'Multiple discriminant function analysis of sex and race in the postcranial skeleton' *American Journal of Physical Anthropology* 61: 305–14

Dittrick, J., Suchey, J.M. 1986: 'Sex determination of prehistoric central California skeletal remains using discriminant analysis of the femur and humerus' *American Journal of Physical Anthropology* 70: 3–9

Divakar, D.D., John, J., Al Kheraif, A.A., Mavinapalla, S., Ramakrishnaiah, R., Vellappally, S., Hashem, M.İ., Dalati, M.H.N., Durgesh, B.H., Safadi, R.A., Anil, S. 2016: 'Sex determination using discriminant function analysis in Indigenous (Kurubas) children and adolescents of Coorg, Karnataka, India: A lateral cephalometric study' *Saudi Journal of Biological Sciences* 23.6: 782–88

Durić, M., Racocević, Z., Donić, D. 2005: 'The reliability of sex determination of skeletons from forensic context in the Balkans' *Forensic Science International* 147: 159–64

Erdal, Y.S. 2008: 'Occlusal grooves in anterior dentition among Kovuklukaya inhabitants (Sinop, Northern Anatolia, 10th Century AD)' *International Journal of Osteoarchaeology* 18: 152–66

— 2010: 'İkiztepe yerleşimi ışığında Orta Karadeniz Bölgesi insanlarının antropolojik yapısı' *Anadolu Araştırmaları* 19.1: 67–97

Günay, Y., Altınkök, M. 2000: 'The value of the size of foramen magnum in sex determination' *Journal of Clinical Forensic Medicine* 7.3:147–49

Hanihara, K. 1958: 'Sexual diagnosis of Japanese long bones by means of discriminant function' *Journal of the Anthropological Society of Nippon* 66: 187–96

— 1959: 'Sex diagnosis of Japanese skulls and scapulae by means of discriminant functions' *Journal of the Anthropological Society of Nippon* 67: 191–97

— 1981: 'Sexing of Japanese skeleton teeth by discriminant function method' *Journal of Anthropological Society of Nippon* 89: 401–18

Holland, T.D. 1991: 'Sex assessment using the proximal tibia' *American Journal of Physical Anthropology* 85: 221–27

Holman, D.J., Bennett, K.A. 1991: 'Determination of sex from arm bone measurements' *American Journal of Physical Anthropology* 84: 421–26

Inoue, M., Inoue, T., Fushimi, Y., Okada, K. 1992: 'Sex determination by discriminant function analysis of lateral cranial form' *Forensic Science International* 57: 109–17

İşcan, M.Y., Ding, S. 1995: 'Sexual dimorphism in the Chinese femur' *Forensic Science International* 74: 79–87

İşcan, M.Y., Loth, S.R., King, C.A., Shihai, D., Yoshino, M. 1998: 'Sexual dimorphism in the humerus: a comparative analysis of Chinese, Japanese and Thais' *Forensic Science International* 98: 17–29

İşcan, M.Y., Miller-Shaivitz, P. 1984: 'Determination of sex from the tibia' *American Journal of Physical Anthropology* 64: 53–57

— 1986: 'Sexual dimorphism in the femur and tibia' in K.J. Reichs (ed.), *Forensic Osteology: Advances in the Identification of Human Remains*. Springfield, IL, Charles C. Thomas: 102–11

İşcan, M.Y., Steyn, M. 2013: *The Human Skeleton in Forensic Medicine*, 3rd Edition. Springfield, IL, Charles C. Thomas Publisher

İşcan, M.Y., Yoshino, M., Kato, S. 1994: 'Sex determination from the tibia: standards for contemporary Japan' *Journal of Forensic Sciences* 39: 785–92

— 1995: 'Sexual dimorphism in modem Japanese crania' *American Journal of Human Biology* 7: 459–64

Karaman, F. 2006: 'Use of diagonal teeth measurements in predicting gender in a Turkish population' *Journal of Forensic Science* 51.3: 630–35

King, C.A., İşcan, M.Y., Loth, S.R. 1998: 'Metric and comparative analysis of sexual dimorphism in the Thai femur' *Journal of Forensic Science* 43: 954–58

Koca Özer, B., Özer, İ., Sağır, M., Güleç, E. 2014: 'Sex determination using tibia in an ancient Anatolian population' *Mediterranean Archaeology and Archaeometry* 14.2: 329–36

Krogman, W.M. 1939: 'A guide to the identification of human skeletal material' *FBI Law Enforcement Bulletin* 8.8: 1–29

Li, C. 1991: 'Sex determination by skull' *Journal of Chinese Forensic Medicine* 6: 120

Liu, W. 1989: 'Sex discriminant analysis of long bones of upper limb' *Acta Anthropologica Sinica* 8: 231–39

Liu, W., Yang, M., Tai, F. 1989: 'Sex discriminant analysis of long bones of lower limb' *Acta Anthropologica Sinica* 8: 147–54

MacLaughlin, S.M., Bruce, M.F. 1985: 'A simple univariate technique for determining sex from fragmentary femora: its application to a Scottish short cist population' *American Journal of Physical Anthropology* 67: 413–17

Mall, G., Graw, M., Gehring, K.D., Hubig, M. 2000: 'Determination of sex from femora' *Forensic Science International* 113.1–3: 315–21

Nakahashi, T., Nagai, M. 1986: 'Sex assessment of fragmentary skeletal remains' *Journal of the Anthropological Society of Nippon* 94: 289–305

Mall, G., Hubig, M., Buttner, A., Kuznik, J., Penning, R., Graw, M. 2001: 'Sex determination and estimation of stature from the long bones of the arm' *Forensic Science International* 117: 23–30

Milner, G.R., Wood, J.W., Boldsen, J.L. 2019: 'Paleodemography: problems, progress, and potential' in M.A. Katzenberg, A.L. Grauer (eds) *Biological Anthropology of the Human Skeleton*, 3rd Edition. Hoboken, Wiley Blackwell: 593–633

Özer, İ. 1999: *Dilkaya (Van) popülasyonunun diskriminant fonksiyon analizi ve Anadolu toplulukları arasındaki yeri.* PhD thesis, Ankara Üniversitesi

— 2014: 'Eski Anadolu ve Japon iskeletlerinde diskriminant fonksiyon analiziyle cinsiyet tayini' *OLBA* 22: 1–13

Özer, İ., Katayama, K. 2006: 'Sex determination using the femur in an ancient Anatolian population' *Anthropologischer Anzeiger* 64.4: 389–98

— 2008: 'Sex determination using the femur in an ancient Japanese population' *Collegium Anthropologicum* 32.1: 67–72

Özer, İ., Katayama, K., Sağır, M., Güleç, E. 2006: 'Sex determination using the scapula in medieval skeletons from East Anatolia' *Collegium Anthropologicum* 30.2: 415–19

Özer, İ., Sağır, M. 2002: 'Diskriminant analiz yardımıyla kafatasından cinsiyet tayini' *Antropoloji Dergisi* 14: 115–26

— 2003: 'Dilkaya ve bazı eski Anadolu toplumlarında uzun kemiklerin çevre ölçülerinden cinsiyet tayini' *Antropoloji Dergisi* 16: 87–98

— 2004: 'Osmanlı Kafatası Endislerinde Discriminant Analiziyle Cinsiyet Tayini' *Antropoloji Dergisi* 19: 53–66

Pandya, A.M., Singel, T.C., Akbari, V.J., Dangar, K.P., Tank, K.C., Patel, M.P. 2011: 'Sexual dimorphism of maximum femoral length' *National Journal of Medical Research* 1.1: 12–31

Purkait, R. 2001: 'Measurements of ulna-a new method for determination of sex' *Journal of Forensic Sciences* 46: 924–27

— 2003: 'Sex determination from femoral head measurement: a new approach' *Legal Medicine* 5: S347–S350

Purkait, R., Chandra, H. 2002: 'Sexual dimorphism in femora: an Indian study' *Forensic Science Communications* 4.3: 1–6

— 2004: 'A study of sexual variation in Indian femur' *Forensic Science International* 146: 25–33

Ren, G. 1987: 'Sexual diagnosis of scapula by discriminant funtion analysis' *Acta Anthropologica Sinica* 6.2: 144–46

Sakaue, K. 2004: 'Sexual determination of long bones in recent Japanese' *Antropological Science* 112: 75–81

Saunders, S. 1992: 'Subadult skeletons and growth studies' in S.R. Saunders, M.A. Katzenberg (eds) *Skeletal Biology of Past Peoples: Research Methods*. New York, Wiley-Liss: 1–20

Schulter-Ellis, F.P., Hayek, L.C., Schmidt, D.J. 1985: 'Determination of sex with discriminant analysis of new pelvic bone measurements: Part II' *Journal of Forensic Science* 30.1: 178–85

Schulter-Ellis, F.P., Schmidt, D.J., Hayek, L.A., Craig, J. 1983: 'Determination of sex with a discriminant analysis of new pelvic bone measurements: Part I' *Journal of Forensic Science* 28.1: 169–80

Singh, G., Singh, S., Singh, S.P. 1975: 'Identification of sex from the tibia' *Journal of Anatomical Society of India* 24: 20–24

Song, H.W., Lin, Z.Q., Jia, J.T. 1992: 'Sex diagnosis of Chinese skulls using multiple stepwise discriminant function analysis' *Forensic Science International* 54: 135–40

Soni, G., Dhall, U., Chhabra, S. 2010: 'Determination of sex from femur: discriminant analysis' *Journal of Anatomical Society of India* 59: 216–21

Spradley, M.K., Jantz, R.L. 2011: 'Sex estimation in forensic anthropology: skull versus postcranial elements' *Journal of Forensic Science* 56.2: 289–96

Stewart, T.D. 1948: 'Medico-legal aspects of the skeleton: age, sex, race and stature' *American Journal of Physical Anthropology* 6: 315–21

— 1951: 'What the bones tell' *FBI Law Enforcement Bulletin* 20.2: 2–5

Steyn, M., İşcan, M.Y. 1997: 'Sex determination from the femur and tibia in South African whites' *Forensic Science International* 90: 111–19

Tanaka, T., Hanihara, K., Koizumu, K. 1979: 'Sex determination of the modern Japanese skull by means of discriminant function' *Sapporo Medical Journal* 48: 583–93

Taylor, J.V., DiBennardo, R. 1982: 'Determination of sex white femora by discriminant function analysis: forensic science applications' *Journal of Forensic Sciences* 27: 417–23

Trancho, G.J., Robledo, B., Lopez-Bueis, I., Sanchez, S.A. 1997: 'Sexual determination of femur using discriminate function analysis of a Spanish population of known sex and age' *Journal of Forensic Sciences* 42: 181–85

Usta, E. 2019: *Eski Anadolu topluluklarında metrik yöntemlerle cinsiyet tayini: Dilkaya örneği.* Master's thesis, Ankara Üniversitesi

Vatandaş, C. 2007: 'Toplumsal cinsiyet ve cinsiyet rollerinin algılanışı' *Sosyoloji Konferansları* 35: 29–56

Walker, P.L., Cook, D.C. 1998: 'Gender and sex: vive la difference' *American Journal of Physical Anthropology* 106: 255–59

White, T.D., Black, M.T., Folkens, P.A. 2012: *Human Osteology*, 3rd Edition. San Diego, Elsevier

Yang, Z.P., Yang, S.F., Lin, Y.C., Shay, J.C., Chi, C.Y. 1988: 'C-shaped root canals in mandibular second molars in a Chinese population' *Endodontics and Dental Traumatology* 4: 160–63

Zheng, T., Pang, T. 1988: 'Determination of sex from the tibia by the stepwise discriminatory analysis' *Acta Anthropologica Sinica* 7.2: 154–59

3. Stature estimation and stature changes in Anatolia

Mehmet Sağır

Introduction

The physical stature of people differs in proportion to the growth and development characteristics of a society. In addition to genetics, stature is affected by numerous factors with regard to time and place. These include, for example, nutrition and environment; children and adolescents who are nutritionally deprived or who suffer from serious diseases might become shorter adults. Therefore, stature can be an important sign in monitoring disease or malnutrition, and for interpreting weight (Lohman et al. 1988; Güleç 1990).

Stature is an important indicator of a person's physical characteristics, including overall body size and bone length. An increase in stature is associated with general morphological development over the lifetime of human beings. Growth is defined as an increase in the numbers and mass of body cells, and the development of bones and cartilage. A positive metabolic accumulation of water, nitrogen and electrolytes is also seen when stature and weight increase during growth. An increase in stature is significantly affected by an individual's nutrition and the essential items for growth obtained through their diet. Individuals of higher socioeconomic status have been found to grow more rapidly and in a more obvious manner. Research shows that socioeconomic status is associated with a wide array of health, cognitive and socioemotional outcomes in children, with effects beginning before birth and continuing into adulthood. Factors affecting individual characteristics during growth are classified into two groups. The first are the factors inherited from the family (i.e. genetic material) and the second are environmental factors such as nutrition, climate and socioeconomic level (Bradley, Corwyn 2002).

Physical characteristics are defining traits or features of a person's body. These are aspects of appearance that are usually outwardly visible to others, even in the absence of other information about the person. Stature is the first thing to be identified when describing the physical characteristics of a person. Therefore, scientists working on this subject classify stature within certain limits based on general prevalence percentages in societies. Martin (1914) and Vandervael (1943) determined certain classifications with regard to this subject (Tab. 3.1) (Güleç 1990).

Measuring and analysing stature in order to define the physical characteristics of a community can easily be done on living people. Stature can be measured by using anthropometry methods; however, it is not always possible to measure the stature of skeletons revealed during excavations of archaeological sites or encountered in forensic investigations/contexts. Skeletons from these types of postmortem contexts are often incomplete and sometimes represented by only a single skeletal element, such as a long bone. These bones may help to obtain information about the physical structures of the people

Classes	*Male*	*Female*	*Male*	*Female*
Midget	x–130	–	x–125	–
Very Short	130–149.9	121–139.9	125–155	x–147.5
Short	150–159.9	140–148.9	155–161	147.5–152.5
Below Average	160–163.9	149–152.9	161.5–167.5	153–158
Average	164–166.9	153–155.9	168.174	158.5–163.5
Above Average	167–169.9	156–158.9	174.5–180.5	164–169
Tall	170–179.9	159–167.9	181–187	169.5–174.5
Very Tall	180–199.9	168–186.9	187–200	174.5+
Giant	200–x	187–x	200–x	200–x

Tab. 3.1. Martin's and Vandervael's stature classifications (Martin 1914; Vandervael 1943).

from those ancient settlements and the identification of a skeleton as an individual. One of the most important aspects that will give us information about these structures is stature.

The methods used for estimating and calculating the stature of skeletons involve examining the long bones, including the femur, tibia, fibula, humerus, radius and ulna. The femur, tibia and vertebrae are known to be directly correlated with stature. Bones such as the clavicula, metatarsal, metacarpal and vertebrae are also used for estimating stature. Numerous studies have been conducted on many communities, and different stature calculation formulas have been developed for different societies. Since the earliest studies in the 1800s, the methods to determine stature have mainly concentrated on forensic medicine and physical anthropology. These studies have developed in two directions: one is the 'anatomical method', based on complete or relatively intact skeletal structures. The other is the 'mathematical method', which is used in the case of incomplete, unarticulated or disturbed skeletons, and utilises mathematical equations to calculate a linear relationship between long bones and stature (Lundy 1987; Raxter et al. 2006; Bidmos, Mazengenya 2020).

The stature of skeletons from archaeological excavations in Anatolia, as well as the stature of modern skeletons, is calculated using formulas developed following research conducted on different societies and based on established methods. As these formulas are developed following analysis of particular and distinct study populations, a population's morphological characteristics, which form the essence of the formula, must be similar to the morphological characteristics of the skeletal population being examined (Trotter, Gleser 1958; Boldsen 1990).

Estimating stature from long bones is important as a part of identifying unknown human remains. Certain issues must be clarified when human skeletons are unearthed during archaeological excavations or found in forensic contexts. Firstly, the identity of the skeleton must be defined; therefore, the sex, age-at-death and physical characteristics of the skeleton must be determined – including an estimation of stature. The two main methods used to estimate stature, as mentioned, are the 'mathematical method' and the 'anatomical method'; the latter is based on re-articulating the complete skeleton with the vertebral column, and including the (estimated presence of the) soft tissues as well (Telkka 1950; Musgrave, Harneja 1978; Lundy 1987; Boldsen 1990).

The anatomical method

The 'anatomical method' was first used by Dwight in the 1890s. In this method, the bones forming the skeleton are laid out in their anatomical position and assembled from the joints, and then the length of the skeleton is measured (Dwight 1894). Dwight's method was updated by Fully in 1956, whose method involved arranging the skeleton in anatomical position and measuring the bones separately before adding in the estimated thickness of the soft tissues. The basio-bregmatic height of the skull (as defined by Hrdlicka 1939), the greatest anterior height of all of the vertebrae from the first cervical vertebra to the first sacral vertebra, the bi-condylar length of the femur, the length of the tibia without spines, and the length of the talus and calcaneus in normal anatomical position are the measurements included in Fully's (1956) method. The advantage of Fully's method when compared to Dwight's is that it is not necessary to fully build and articulate the skeleton. For this reason, Fully's method is more useful than Dwight's in both laboratory and field

Regression formula for males	*Regression formula for females*
Stature = 81.306 + 1.880 * F	Stature = 72.844 + 1.945 * F
Stature = 70.641 + 2.894 * H	Stature = 71.475 + 2.754 * H
Stature = 78.664 + 2.378 * T	Stature = 74.774 + 2.352 * T
Stature = 85.925 + 3.271 * R	Stature = 81.224 + 3.343 * R
Stature = 71.272 + 1.159 * (F + T)	Stature = 69.154 + 1.126 * (F + T)
Stature = 71.441 + 1.220 * F + 1.080 * T	Stature = 69.561 + 1.117 * F + 1.125 * T
Stature = 66.855 + 1.730 * (H + R)	Stature = 69.911 + 1.628 * (H + R)
Stature = 69.788 + 2.769 * H + 0.195 * R	Stature = 70.542 + 2.582 * H + 0.281 * R
Stature = 68.397 + 1.030 * F + 1.557 * H	Stature = 67.435 + 1.339 * F + 1.027 * H
Stature = 67.049 + 0.913 * F + 0.600 * T + 1.225 * H - 0.187 * R	Stature = 67.467 + 0.782 * F + 1.120 * T +1.059 * H - 0.711 * R

Tab. 3.2. Regression formulas established by Pearson (1899) for estimating stature of females and males.

settings (Lundy 1987). Olivier and Pineau (1960) developed a slightly different method that estimates height based on measurements of the tibia and the five lumbar vertebrae (Dwight 1894; Hrdlicka 1939; Olivier, Pineau 1960; Olivier 1969).

The mathematical method

The 'mathematical method' is the most commonly utilised and applied method today, as it does not require a complete cadaver or skeleton, and its implementation is relatively simple. In this method, the height of the individual is estimated by establishing the linear relationship between a single long bone or a combination of several long bones, and comparing the calculated proportions with measurements from living individuals or cadavers. It has been suggested that the most appropriate bones for estimating stature with the mathematical method are the femur and tibia, which have been shown to have a direct relationship with height (Lundy 1987; Aiello, Dean 1990; Holland 1995).

Historical background to stature estimation

Numerous formulas have been developed and established over time by researchers measuring long bones and applying mathematical methods. In these formulas, the length of an individual is estimated using the greatest length of the long bones. The first study on this subject was conducted by Rollet in 1888. Rollet used the humerus, radius, ulna, femur, tibia and fibula bones from 50 male and 50 female French cadavers, and established the oldest formulated height table. In the 1890s, Manouvrier re-evaluated Rollet's data. Rollet excluded individuals older than 60 years old, and his study was conducted on 26 males and 25 females. Manouvrier noted that when individuals over 60 were included, a loss of approximately 3cm would be observed in stature calculations (Manouvrier 1893; Olivier 1969; Olivier et al. 1978; Krogman, İşcan 1986). In 1899, Pearson established the first regression formulas using only the right-side long bones, following Rollet's data (Tab. 3.2). Pearson stated that attention must be paid while applying these formulas to a different population group, and emphasised that the stature decrease observed in individuals older than 60 years old, as suggested by Manouvrier, is not of great significance. In 1929, Stevenson compared the regression formulas he obtained through his studies on communities in northern China with Rollet's data on French cadavers, and affirmed Pearson's hypothesis that 'attention must be paid while applying stature regression formulas on another population group' (Stevenson 1929; Krogman, İşcan 1986).

Anthropologists aiming to establish certain basic principles for stature estimation studies came together at an international meeting in Geneva, Switzerland in 1912, and expressed their opinions on stature, ultimately agreeing on a method for measuring long bones. It was decided to measure stature by using the greatest lengths of the long bones, but it was emphasised that the spina must be excluded while taking the maximum length of the tibia (Maccurdy 1912). Telkka (1950) created tables and regression formulas for determining the stature of 115 Finnish skeletons, and he emphasised that there was an asymmetry between the right and left sides of the long bones in these studies, and included these right-left differences in his tables. In 1951, Dupertuis and Hadden also created formulas for determining the height of black and white people in North America, including 100 individuals from each sex and ethnic group housed in the

Males	*Females*
69.089 + 2.236 Fem	61.412 + 2.317 Fem
81.688 + 2.392 Tib	72.572 + 2.533 Tib
73.570 + 2.970 Hum	64.977 + 3.144 Hum
80.405 + 3.650 Rad	73.502 + 3.876 Rad
69.294 + 1.225 (Fem + Tib)	65.213 + 1.233 (Fem + Tib)
71.429 + 1.728 (Hum + Rad)	55.729 + 1.984 (Hum + Rad)
66.544 + 1.422 Fem + 1.062 Tib	59.259 + 1.657 Fem + 0.879 Tib
66.400 + 1.789 Hum + 1.841 Rad	60.344 + 2.164 Hum + 1.525 Rad
64.505 + 1.928 Fem + 0.568 Hum	57.600 + 2.009 Fem + 0.566 Hum
78.772 + 2.102 Tib + 0.606 Rad	65.354 + 2.082 Tib + 0.689 Rad
56.006 + 1.442 Fem + 0.931 Tib + 0.083 Hum + 0.480 Rad	57.495 + 1.544 Fem + 0.764 Tib + 0.126 Hum + 0.295 Rad

Tab. 3.3. Stature regression formulas established by Dupertuis and Hadden (1951).

Hamann-Todd cadaver collection. They were of the opinion that the difference between antemortem stature and the cadaver stature did not have a significant effect on the measurement results, and emphasised that this difference could be ignored (Tab. 3.3) (Dupertius, Hadden 1951; Krogman, İşcan 1986).

Trotter and Gleser, who are now well known for their work on stature estimation, published their first study in 1952. Studying a collection of skeletal material comprised of the bones of US soldiers who died in the Second World War and from the Terry Collection, Trotter and Gleser had the opportunity to study the bones of people whose stature was known before they died. They studied all six long bones of the human skeletal system and created regression formulas for estimating the stature of black and white North Americans (Tab. 3.4) (Trotter, Gleser 1952).

In 1956, Fully was assigned to identify the remains of French skeletons found at a concentration camp. He determined the stature of these skeletons by applying Manouvrier's formulas; later, records of the stature of some of the living individuals were found in the archives, and it was revealed that there were significant deviations from the statures calculated using Manouvrier's formulas. Following this, Fully developed a different estimation method, taking into account body proportions in order to overcome this problem. He then calculated stature, including the basion-bregma height of the skull, the length of the vertebra from the second cervical to the fifth lumbar and the length of the first sacral vertebra, as well as the lengths of the femur, tibia, calcaneus and talus. He emphasised that 10–11.5cm must be added to the stature calculated with this method. Some researchers stated that although the error rate of this method is low, it can only be used in forensic cases where there is a high probability of finding an entire skeleton. In response, Olivier, together with Pineau, developed two updated methods in 1960 (Olivier, Pineau 1960; Krogman, İşcan 1986; Lundy 1987).

Trotter and Gleser had the opportunity to conduct a comprehensive study in 1958 on the skeletons of US soldiers who had died in the Korean War. In this study, they tested their previous work and also developed new

White males	*S.E.*	*Black males*	*S.E.*
3.08 Hum + 70.45	4.05	3.26 Hum + 62.10	4.43
3.78 Rad + 79.01	4.32	3.42 Rad + 81.56	4.30
3.70 Ulna + 74.05	4.32	3.26 Ulna + 79.29	4.42
2.38 Fem + 61.41	3.27	2.11 Fem + 70.35	3.94
2.52 Tib + 78.62	3.37	2.19 Tib + 86.02	3.78
2.68 Fib + 71.78	3.29	2.19 Fib + 85.65	4.08
1.30 (Fem + Tib) + 63.29	2.99	1.15 (Fem + Tib) + 71.04	3.53
1.42 Fem + 1.24 Tib + 59.88	2.99	0.66 Fem + 1.62 Tib + 76.13	3.49
0.93 Hum + 1.94 Tib + 69.30	3.26	0.90 Hum 1.78 Tib + 71.29	3.49
0.27 Hum + 1.32 Fem + 1.16 Tib + 58.57	3.51	0.89 Hum - 1.01 Rad + 0.38 Fem + 1.92 Tib + 74.56	3.22

White females	*S.E.*	*Black females*	*S.E.*
3.36 Hum + 57.97	4.45	3.08 Hum + 64.67	4.25
4.74 Rad + 54.93	4.24	3.67 Rad + 71.79	4.59
4.27 Ulna + 57.76	4.30	3.31 Ulna + 75.38	4.83
2.47 Fem + 54.10	3.72	2.28 Fem + 59.76	3.41
2.90 Tib + 61.53	3.66	2.45 Tib + 72.65	3.70
2.93 Fib + 59.61	3.57	2.49 Fib + 70.90	3.80
1.39 (Fem + Tib) + 53.20	3.55	1.26 (Fem + Tib) + 59.72	3.28
1.48 Fem + 1.28 Tib + 53.07	3.55	1.53 Fem + 0.96 Tib + 58.54	3.23
1.35 Hum + 1.95 Tib + 52.77	3.67	1.08 Hum + 1.79 Tib + 62.80	3.58
0.68 Hum + 1.17 Fem + 1.15 Tib + 50.122	3.51	0.44 Hum - 0.20 Rad + 1.46 Fem + 0.86 Tib + 56.33	3.22

Tab. 3.4. Stature regression formulas and standard errors for white and black North Americans (Trotter, Gleser 1952).

White	*S.E.*	*Black*	*S.E.*
1.31 (Fem + Fib) + 63.05	3.62	1.20 (Fem + Fib) + 67.77	3.63
1.26 (Fem + Tib) + 67.09	3.74	1.15 (Fem + Tib) + 71.75	3.68
2.60 Fib +75.50	3.86	2.10 Fem + 72.22	3.91
2.32 Fem + 65.53	3.94	2.19 Tib + 85.36	3.96
2.42 Tib + 81.83	3.37	2.34 Fib + 80.07	4.02
1.82 (Hum + Rad) + 67.97	2.99	1.66 (Hum + Rad) + 73.08	4.18
1.78 (Hum + Ulna) + 66.98	2.99	1.65 (Hum + Ulna) + 70.67	4.23
2.89 Hum + 78.10	4.05	2.88 Hum + 75.48	4.23
3.79 Rad + 79.42	4.32	3.32 Rad + 85.43	4.57
3.76 Ulna + 75.55	4.32	3.20 Ulna + 82.77	4.74
Mongolian	*S.E.*	*Mexican*	*S.E.*
1.22 (Fem + Fib) + 70.24	3.18		
1.22 (Fem + Tib) + 70.37	3.24		
2.40 Fib + 80.56	3.24	2.40 Fem + 58.67	2.99
2.39 Tib + 81.45	3.27	2.50 Fib + 75.44	3.52
2.15 Fem + 72.57	3.80	2.34 Tib + 80.07	3.73
1.68 (Hum + Ulna) + 71.18	4.14		
1.67 (Hum + Rad) + 74.83	4.16		
2.68 Hum + 83.19	4.25	3.55 Rad + 80.71	4.04
3.54 Rad + 82.00	4.60	3.56 Ulna + 74.56	4.05
3.48 Ulna + 77.45	4.66	2.92 Hum + 73.94	4.24

Tab. 3.5. Regression formulas created by Trotter and Gleser for white and black North American and Mongolian and Mexican males (Trotter, Gleser 1958).

and more comprehensive regression formulas by examining a large number of black and white people, in addition to a small number of Mongolian, Mexican and Puerto Rican skeletons (Tab. 3.5). Trotter and Gleser studied a total of 5,517 cadavers aged between 17 and 46. Based on their regression formulas, they suggested that one or a combination of the formulas developed by other researchers for work on populations of different 'races', geographies and generations must be implemented in each case (Trotter, Gleser 1958).

Olivier et al. (1978) tried to develop new regression formulas for French populations, using both Fully's and Manouvrier's data. However, due to insufficient statistical data and the low number of individuals in Manouvrier's tables, they could not use the data obtained from these previous studies but instead developed new regression formulas based only on Fully's data. Olivier et al. (1978) stated that the estimations based on the leg bones gave better results than the arms, and the error rate of the results obtained from the combination of the femur and other bones was reduced. In 1967, Genoves created new regression formulas for use in stature estimation as a result of his study conducted on Mexicans, which used the average lengths of the right and left leg bones from cadavers of people from mostly lower socioeconomic strata (Tab. 3.6) (Genoves 1967; Ubelaker 1978; Krogman, İşcan 1986).

In addition to the studies which have been described so far, numerous other studies have been conducted by different researchers in many parts of the world to estimate

Males	*S.E.*	*Females*	*S.E.*
Stature = 2.26 Fem + 66.379	3.417	Stature = 2.59 Fem + 49.742	3.816
Stature = 1.96 Tib + 93.752	2.812	Stature = 2.72 Tib + 63.781	3.513

Tab. 3.6. Stature regression formulas created by Genoves (1967).

Males	*S.E.*	*Females*	*S.E.*
Stature = 2.441 * Ulna + 105.72	4.64	Stature = 4.191 * Ulna + 58.20	3.99
Stature = 3.128 * Radius + 93.18	4.69	Stature = 4.068 * Radius + 67.19	4.31
Stature = 1.487 * Femur + 101.2	3.74	Stature = 2.754 * Femur + 42.81	5.33
Stature = 1.943 * Tibia + 91.57	3.85	Stature = 3.061 * Tibia + 44.63	4.30
Stature = 2.578 * Fibula + 72.34	4.21	Stature = 3.200 * Fibula + 42.96	5.29

Tab. 3.7. Regression formulas created by Sağır (1994) using X-Ray films.

stature. The ones mentioned above are the most commonly used and discussed studies. Apart from these, there are also other studies like the ones conducted by Pan in India in 1924, by Nat in India in 1931 and by Breitinger in southern Germany in 1937. In 1953, Lorke et al. conducted studies to estimate stature using the right long bones of cadavers between the ages of 19 and 60 years old. In 1961, Allbrock examined the leg-stature correlations of African and British populations aged 18–34 and 20–22 years old. In 1966, Eliakis et al. published the equations they obtained from a Greek population aged 22–65 years old. Dayala et al. (2008) published the equations for South African whites. In 2011, Mahakkanukrauh et al. conducted studies to estimate stature using long bones from cadavers of Thai populations.

Apart from the bones used in the studies mentioned above, some other studies have also been conducted using different bones. Musgrave and Harneja (1978) created regression formulas by studying the five metacarpals of 120 males and 46 females aged 17–87 years old living in England. Holland (1995) created regression formulas by studying 119 calcaneus and talus bones belonging to people aged 16–81 years old from the Hamann-Todd collection.

There have also been studies conducted for estimating height using fragmented long bones. In cases where the long bones were fragmented or broken, some researchers developed new formulas using these fragments due to the previous inability to estimate height from these bones. Müller (1935) conducted one such study. Steel and McKern developed formulas in the late 1960s and 1970s for calculating the total lengths of bones based on some anatomical points, from which they then calculated the lengths by applying the total lengths obtained from these formulas to the formulas created by Trotter and Gleser (Steel, McKern 1969). Other similar studies include those conducted by Chandra et al. in 1966, by Mysorekar et al. in 1980, and by Badkur and Nath in 1990.

Studies have also been conducted to estimate the stature of children. These include Telkka et al. (1962), who studied the radiographs of 3848 long bones of children younger than 15 years old, as well as Zorab et al. (1964), who studied subadult tibias, and Feldesman (1992). Such studies have increased in number and scope for different populations around the world, and new results are still being revealed (Telkka et al. 1962; Zorab et al. 1964; Feldesman 1992).

Stature estimation studies in Türkiye

In 1992, Seyhan conducted a study aiming to calculate an individual's height based on spine length. In 1994, Sağır created regression formulas by studying the ulna, radius, femur, tibia and fibula using X-rays of these bones from 166 males and 133 females. The average age-at-death of these individuals was found to be 34.81 for males and 32.34 for females (Tab. 3.7).

In 1994, Şam developed stature regression formulas using fibula lengths in his study conducted on 82 cadavers (51 males and 31 females) aged 16–63 years old (Tab. 3.8).

Günay (1995) also developed stature regression formulas for the tibia by studying 49 male and 31 female cadavers aged 18–54 years old (Tab. 3.9).

Males	*S.E.*
Stature = 2.91 Fibula + 63.73	4.43
Females	*S.E.*
Stature = 2.12 Fibula + 85.64	3.89

Tab. 3.8. Regression formulas created by Şam (1994).

Males	*S.E.*
Stature = 2.7 Tibia + 71.2	4.8
Females	*S.E.*
Stature = 1.9 Tibia + 92.4	4.0

Tab. 3.9. Regression formulas created by Günay (1995).

In his comprehensive study, Sağır (2000) created regression equations including all bones and their combinations for 100 females and 100 males (between the ages of 18 and 30 years old) through an examination of the bones from X-rays. The equations created in this study are used for determining the height of archaeological skeletons and also in forensic medicine (Tab. 3.10). Sağır (2006) also developed regression formulas for determining the heights of males and females using the length of all metacarpal bones from a study population of 100 females and 100 males.

More recently, many regression analyses have been developed in forensic medicine and forensic anthropology in order to estimate stature, and several formulas have been created from the bones of different parts of the body. Karaçam et al. (2005) performed length calculations for the femur using the femur bones of cadavers. In 2006, Celbiş and Ağrıtmış developed regression equations to be used for calculating stature following analysis of the ulna bones of 80 male and 47 female cadavers. In 2007, Karaman et al. created regression equations in order to calculate stature from hand sizes and metacarpals following analysis of 46 females and 56 males using X-rays. Regression analyses were performed in order to determine stature from the second cervical vertebra following analysis of 320 individuals using magnetic resonance (Ulu et al. 2020). Studies like these continue to be performed today in increasing numbers, and on different bones of the human skeleton.

Stature change in Anatolia

Türkiye, and thereby Anatolia, has been a location of intense human mobility over the course of history. We know that the journey of hominins, and subsequently modern humans, in this region is more than 1 million years old, from at least the Palaeolithic Age. Archaeological remains have demonstrated that many societies and civilisations have survived in this vast geography during this long period. Of course, thousands of skeletons belonging to the people who created these cultures and civilisations have also been brought to light, along with the architectural and material artefacts and remains. In the analysis of these skeletons, estimations of the stature of these people was attempted using the stature calculation formulas available at the time. In these studies, the formulas of Trotter and Gleser (1952), and especially Pearson (1899) were used.

The heights and statures of past Anatolian populations from the Neolithic to the present have been calculated in these studies. One of the first studies conducted

Male	*Female*
Stature = 2.512 x Humerus + 87.07 ± 4.61	Stature = 2.437 x Humerus + 84.487 ± 3.99
Stature = 3.152 x Radius + 95.403 ± 4.62	Stature = 3.044 x Radius + 94.781 ± 3.53
Stature = 3.310 x Ulna + 85.368 ± 4.52	Stature = 3.111 x Ulna + 87.264 ± 3.63
Stature = 1.941 x Femur + 82.040 ± 4.52	Stature = 1.830 x Femur + 81.158 ± 3.17
Stature = 2.338 x Tibia + 83.329 ± 3.89	Stature = 2,073 x Tibia + 87.696 ± 3.58
Stature = 2.379 x Fibula + 82.75 ± 3.81	Stature = 2,251 x Fibula + 82.568 ± 3.51
Stature = 1.187 x (Femur+Tibia) + 71.840 ± 3.8905	Stature = 1.077 x (Femur+Tibia) + 75.979 ± 3.1330
Stature = 0.610 x Femur + 1.816 x Tibia + 74.752 ± 3.8050	Stature = 1.404 x Femur + 0.659 x Tibia + 76.480 ± 3.1172
Stature = 1.666 x (Humerus+Radius) + 74.988 ± 4.1561	Stature = 1.581 x (Humerus+Radius) + 77.065 ± 3.3913
Stature = 1.505 x Humerus + 1.876 x Radius + 75.287 ± 4.1705	Stature = 1.120 x Humerus + 2.125 x Radius + 79.719 ± 3.3735
Stature = 1.009 x Humerus + 1.760 x Tibia + 70.923 ± 3.6802	Stature = 1.257 x Humerus + 1.404 x Tibia + 72.010 ± 3.603
Stature = 1.390 x Humerus + 1.187 x Femur + 69.837 ± 4.1794	Stature = 0.838 x Humerus + 1.456 x Femur + 71.233 ± 3.0804
Stature = 0.652 x Humerus + 0.792 x Radius + 0.406 x Femur + 1.231 x Tibia + 64.902 ± 3.6213	Stature = 0.222 x Humerus + 1.289 x Radius + 1.035 x Femur + 0.333 x Tibia + 69.219 ± 3.6213
Stature = 0.874 x Humerus + 0.320 x Femur + 1.564 x Tibia + 68.083 ± 3.6731	Stature = 0.712 x Humerus + 1.196 x Femur + 0.489 x Tibia + 69.251 ± 3.0575

Tab. 3.10. Stature regression formulas created for males and females (Sağır 2000).

on this subject, and some of the first data for Anatolian past populations, was performed by Kansu in 1937 on the skeletons from Alacahöyük. In this study, Kansu used Pearson's (1899) formulas and found the average stature calculated from the skeletons of two males to be 153.25cm. In studies conducted up until the 1980s, often the number of analysed skeletal individuals was quite low, and insufficient for accurate and reliable statistical evaluations. With regard to stature calculated utilising the Pearson formula, the highest number of individuals is represented by three females and two males from Chalcolithic Şeyh Höyük (Tunakan 1951; Şenyürek 1955). In Tunakan's (1951) study, the average stature was found to be 149.11cm for females and 164.59cm for males. Şenyürek (1955), on the other hand, calculated the average stature as 150.64cm for females and 165.46cm for males. From the 1980s onwards the number of skeletons examined (and therefore the number of individuals) increased, and some of these later studies include those conducted by Özbek (Late Byzantine İznik) in 1984, Güleç (medieval Dilkaya) in 1986 and Coşkun (Byzantine Topaklı) in 1988. Özbek calculated the average stature of 35 males from his study population as 167.40cm, Coşkun calculated the average stature of 33 females in the studied population as 154.03cm and the average stature of 38 males as 165.55cm, and Güleç calculated the average stature of 34 females from the studied population as 153.04cm and the average stature of 41 males as 163.89cm.

Researchers have generally used the formulas of Trotter and Gleser (1952), together with Pearson's (1899) formulas, and then stature estimates were attempted by taking the average of the obtained results. However, when the results obtained from the formulas were examined, differences between them were noted. For example, Şenyürek (1949) calculated stature from the Kumtepe skeletal sample population using Pearson's (1899) formula, and determined it as being 153.82cm for females and 160.50cm for males, but when the stature was later calculated using Trotter and Gleser's (1952) formulas, it was found to be 157.02cm for females and 165.83cm for males. The difference between the results of the two formulas is 3.2cm for females and 5.33cm for males. Differences have also been noted in other studies; the male stature calculated in a study of the skeletal sample population from Hittite-period Alacahöyük (Kansu 1937) was found to be 153.25cm according to the Pearson's (1899) formula, then later calculated at 162.04cm using Trotter and Gleser's (1952) formulas. In this instance the difference between the two formulas is very large (8.79cm). Sağır (2000) emphasised that Pearson's (1899) formulas should not be used for stature studies on Anatolian populations, with the obtained results being quite a bit lower than the true stature values, and furthermore that Trotter and Gleser's (1952) formulas gave values which were closer to the results obtained from the formulas which have been developed for Turkish populations. Using the regression formulas

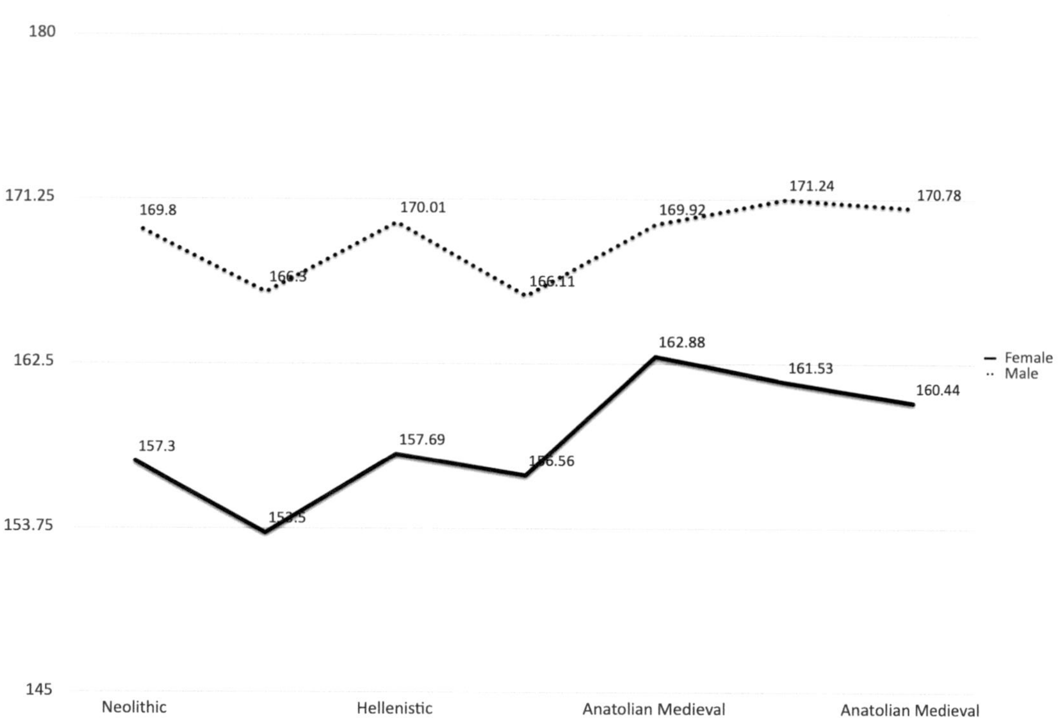

Fig. 3.1. Stature changes in past Anatolian populations (chart by author).

for Turkish populations is very significant to achieve more accurate results in future studies on Turkish or Anatolian material/individuals.

Trotter and Gleser's (1952) formulas have been the most commonly used for evaluating changes in the stature of past Anatolian populations; therefore, in light of these studies, it would be more correct to look at the results obtained from these formulas. When examined chronologically, considering the number of individuals, we can see that stature during the Anatolian Neolithic was 157.30cm for females and 169.80cm for males (Ferembach 1972). During the Early Bronze Age, stature was calculated as 153.50cm for females and 166.30cm for males (Angel 1970). Sağır et al. (2004) calculated stature at 157.69cm for females and 170.01cm for males in a skeletal population from Börükçü belonging to the Hellenistic period. Çiner (1971) examined the stature of numerous Roman-period skeletons excavated at Gordion, where the stature was found to be 156.56cm for females and 166.11cm for males. Coşkun (1988) calculated the stature of Byzantine-period skeletons excavated at Topaklı to be 162.88cm for females and 169.92cm for males. The medieval skeletons from İznik were studied by Erdal (1991), who calculated stature as 161.53cm for females and 171.24cm for males. Güleç (1989b), studied the Islamic-period skeletons from Panaztepe and calculated stature as 160.44cm for females and 170.78cm for males.

It would be more accurate to evaluate the changes in stature of Anatolian populations by considering all of the data together. However, since the number of examined and analysed individuals was quite low in many studies, only the studies with a statistically sufficient number of individuals are examined here. When the change in stature is examined (Fig. 3.1), we see that there is not a dramatic change since the Neolithic period, except for the Early Bronze Age and Roman period. A remarkable decrease in the height (ca 3cm) of both females and males is observed in these two periods. While observing these changes, we need to consider that Anatolia has very different climatic, geographical, environmental and demographic conditions. Overall, there is a general increase in stature in Anatolia for females since the Neolithic, while it has remained at ca 170cm for males. For studies conducted on living Turkish populations, it has been observed that individuals' heights have increased from 152cm to 159cm for females and from 162cm to 174cm for males (Özer et al. 2011).

The majority of the body-height calculation formulas for individuals living in Türkiye have been developed using the mathematical method. Among those developed formulas, the ones which had a large sample size and were based on long bones and their combinations had lower standard errors. Therefore, they have increased reliability and reduced margins of error when applied in practice for estimating stature. Considering the studies carried out on the skeletal series in Anatolia, the formulas of Pearson (of the regression formulas used before the 1990s) produce lower values in comparison to the formulas developed for Türkiye and those developed by Trotter and Gleser. This results in lower average calculated heights for Anatolian individuals and populations. It is therefore advocated that it would not be useful to make use of Pearson's formulas in Türkiye. This also means that the calculation of average body height using Pearson's formulas and Trotter and Gleser's formulas for stature estimation – used in studies before the 1990s, and for comparison after the 1990s – is problematic. The stature estimation values calculated using the body height formulas developed in Türkiye have also demonstrated that, for calculating body height, the long bones of the lower limbs (femur, tibia and fibula) yield more reliable results than the long bones of the upper limbs (Sağır 2000).

Bibliography

Aiello, L., Dean, C. 1990: *An Introduction to Human Evolutionary Anatomy*. Toronto, Academic Press, Harcourt Brace Jovanovich Publishers

Allbrook, D. 1961: 'The estimation of stature in British and East African males. Based on tibial and ulnar bone lengths' *Journal of Forensic Medicine* 8: 15–28

Angel, J.L. 1970: 'Human skeletal remains at Karataş' *American Journal of Archaeology* 74: 253–59

Badkur, P., Nath, S. 1990: 'Use of regression analysis in reconstruction of maximum bone length and living stature from fragmentary measures of the ulna' *Forensic Science International* 45: 15–25

Bidmos, M.A., Mazengenya, P. 2020: 'Accuracies of discriminant function equations for sex estimation using long bones of upper extremities' *International Journal of Legal Medicine* 135: 1095–102

Boldsen, J.L. 1990: 'Body proportion, population structure and height prediction' *Adli Tıp Dergisi* 6: 157–65

Bradley, R.H., Corwyn, R.F. 2002: 'Socioeconomic status and child development' *Annual Review of Psychology* 53: 371–99

Breitinger, E. 1937: 'Zur Berechnung der Körperhöhe aus den langen Gliedmassenknochen' *Anthropologischer Anzeiger* 14: 249–74

Brooks, S., Steele, D.G., Brooks, R.H. 1990: 'Formulae for stature estimation on incomplete long bones: a survey of their reliability' *Adli Tıp Dergisi* 6: 167–70

Celbiş, O., Ağrıtmış H. 2006: 'Estimation of stature and determination of sex from radial and ulnar bone lengths in a Turkish corpse sample' *Forensic Science International* 158: 135–39

Chandra, A., Dass, R., Khanna, J.K. 1966: 'Reconstruction of the total length of femur from its fragments. A preliminary report' *Journal of the Anatomical Society India* 15: 92–95

Coşkun, M. 1988: *Topaklı Populasyonun Boy Açısından İncelenmesi ve Anadolu Kronolojisindeki Yeri*. Master's thesis, Ankara University, Ankara

Çiner, R. 1971: 'Gordion Roma halkı femur ve tibialarının tetkiki' *Antropoloji* 2: 56–81

Dayala, M.R., Steyn, M., Kuykendall, K.L. 2008: 'Stature estimation from bones of South African whites' *South African Journal of Science* 104: 124–28

Dupertius, C.W., Hadden, J.A. 1951: 'On the reconstruction of stature from long bones' *American Journal of Physical Anthropology* 9: 15–53

Dwight, T. 1894: 'Methods of estimating the height from parts of the skeleton' *Medical Record* 46: 293–96

Eliakis, C., Eliakis, E., Iordanidis, P. 1966: 'Détermination de la taille d'après la mensuration des os longs (mémoire original – recherche expérimentale)' *Acta Medicinae Legalis et Socialis* 19.3: 343–61

Erdal, Y.S. 1991: *İznik Geç Bizans Dönemi İskeletlerinin Paleoantropolojik Açıdan İncelenmesi*. Master's thesis, Hacettepe University, Ankara

Feldesman, M.R. 1992: 'Femur/stature ratio and estimates of stature in children' *American Journal of Physical Anthropology* 87: 447–59

Ferembach, D. 1972: 'Les Hommes du gisement néolithique de Çatal Hüyük' *Türk Tarih Kongresi* (1970) Ankara 7: 15–21

Fully, G. 1956: 'Une nouvelle méthode de détermination de la taille' *Annales de Médicine Légale* 36: 266–73

Genoves S. 1967: 'Proportionality of the long bones and their relation to stature among Mesoamericans' *American Journal of Physical Anthropology* 26: 67–77

Güleç, E. 1986: 'Van-Dilkaya iskeletlerinin paleoantropolojik incelenmesi' *Kültür Bakanlığı Araştırma Sonuçları Toplantısı* IV: 369–80

— 1989a: 'Klazomenai iskeletlerinin paleoantropolojik açıdan değerlendirilmesi' *Belleten* 53: 207–08

— 1989b: 'Panaztepe iskeletlerinin paleoantropolojik ve paleopatolojik incelenmesi' *Türk Arkeoloji Dergisi* 28: 73–95

— 1990: 'Paleoantropolojik verilere göre eski Anadolu bireylerinin boy açısından incelenmesi' *V. Arkeometri Sonuçları Toplantısı*: 147–60

Günay, Y. 1995: *Tibia Uzunluğundan Vücut Boy Uzunluğunun Hesaplanması*. İstanbul, T.C. Adalet Bakanlığı, Adli Tıp Kurumu Başkanlığı, Uzmanlık Tezi

Holland, T.D. 1995: 'Brief communication: estimation of adult stature from the calcaneus and talus' *American Journal of Physical Anthropology* 96: 315–20

Hrdlicka A. 1939: *Practical Anthropometry*. Philadelphia, Wistar

Kansu, Ş.A. 1937: 'Alacahöyük'te bulunan iskeletlerin antropolojik tetkiki' *Belleten* 1.1: 180–91

Karaçam, B., Büyük, Y., Şam, B. 2005: 'Prediction of stature from femur length amongst Turkish population: an autopsy study' *Adli Bilimler Dergisi* 4.3: 39–44

Karaman, A.G., Teke, H.Y., Günay, I., Doğan, B., Bilge, Y. 2007: 'Height estimation using anthropometric measurements on X-Rays of wrist and metacarpal bones' *The Internet Journal of Biological Anthropology* 2: 1

Krogman, W.M., İşcan, M.Y. 1986: *The Human Skeleton in Forensic Medicine*, Second Edition. Springfield, IL, Charles C. Thomas

Lohman, T.G., Roche, A.F., Martorell, R. 1988: *Anthropometric Standardization Reference Manual*. Champaign, IL, Human Kinetics Books

Lorke, D., Munzner, H., Walter, E. 1953: 'Zur Rekonstruktion der Körpergröße eines Menschen aus den langen Gliedmaßenknochen'. *Deutsche Zeitschrift für die gesamte gerichtliche Medizin* 42: 189–202

Lundy, J.K. 1987: 'Estimating living stature from skeletal remains' *Adli Tıp Dergisi* 3: 103–10

Maccurdy, G.G. 1912: 'International Congress of Prehistoric Anthropology and Archeology, Geneva (International Anthropometric Agreement)' *American Anthropologist* 14: 621–31

Mahakkanukrauh, P., Khanpetch, P., Prasitwattanseree, S., Vichairat, K., Case, D.T. 2011: 'Stature estimation from long bone lengths in a Thai population' *Forensic Science International* 210.1–3: 279–85

Manouvrier, L. 1893: 'Le determination de la taille d'après les grands os des membres' *Bulletins et Memoires de la Societe d'Anthropologie de Paris* 4.2: 347–402

Martin, R. 1914: *Lehrbuch der Anthropologie in systematischer Darstellung: mit besonderer Berücksichtigung der anthropologischen Methoden; für Studierende, Ärzte und Forschungsreisende*. Jena, G. Fischer

Musgrave, J.H., Harneja, N.K. 1978: 'The estimation of adult stature from metacarpal bone length' *American Journal of Physical Anthropology* 48: 113–20

Müller, G. 1935: 'Zur Bestimmung der Lange beschadigter Extremitatenknochen'. *Anthropologischer Anzeiger* 12: 70–72

Mysorekar, N.L., Verma, P.K., Nandekar, A.N., Sharma, T.S.R. 1980: 'Estimation of stature from parts of bones: lower end of femur and upper end of radius' *Medicine, Science and the Law* 20.4: 283–86

Nat, B.S. 1931: 'Estimation of stature from long bones in Indians of the united provinces: a medico-legal inquiry in anthropometry' *Indian Journal of Medical Research* 18: 1245–53

Olivier, G., Pineau H. 1960: 'Nouvelle détermination de la taille foetale d'après les longueurs diaphysaires des os longs' *Annales de Médecine Légale, Criminologie, Police Scientifique et Toxicologie* 40: 141–44

Olivier, G. 1969: *Practical Anthropology*. Springfield, IL, Charles C. Thomas Publisher

Olivier, G., Aaron, C. Fully, G., Tissier, G. 1978: 'New estimations of stature and cranial capacity in modern man' *Journal of Human Evolution* 7: 513–18

Özbek, M. 1984: 'Roma açıkhava tiyatrosundan çıkarılan Bizans iskeletleri' *HÜ Edebiyat Dergisi* 2.1: 81–89

Özer, B.K., Sağır, M., Özer, İ. 2011: 'Secular changes in the height of the inhabitants of Anatolia (Turkey) from the 10th millennium BC to the 20th century AD' *Economics and Human Biology* 9: 211–19

Özer, K. 1993: *Antropometri: Sporda Morfolojik Planlama*. İstanbul, Kazancı Matbaacılık Sanayii A.Ş.

Pan, N. 1924: 'Length of long bones and their proportion to body height in Hindus' *Journal of Anatomy* 58: 374–78

Pearson, K. 1899: 'Mathematical contributions to the theory of evolution. V. On the reconstruction of the stature of prehistoric races' *Philosophical Transactions of the Royal Society London* 192: 169–244

Raxter, M.H., Auerbach, B.M., Christopher, B. Ruff, C.B. 2006: 'Revision of the Fully Technique for estimating statures' *American Journal of Physical Anthropology* 130: 374–84

Rollet, E. 1888: *De la mensuration des os longs des membres dans ses rapports avec l'anthropologie, la clinique et la médecine judiciaire*. MD Thesis, The Faculty of Medicine and Pharmacy of Lyon, Lyon

Sağır, M. 1994: *Uzun Kemik Radyografilerinden Boy Hesaplama Formüllerinin Oluşturulması*. Master's thesis, Ankara University, Ankara

— 2000: *Uzun Kemik Radyografilerinden Boy Formüllerinin Oluşturulması*. PhD thesis, Ankara University, Ankara

— 2006: 'Estimation stature from X-Rays of metacarpals in the Turkish population' *Anthropologisher Anzeiger* 64.4: 377–88

Sağır, M., Özer, İ., Satar Z., Güleç, E. 2004: 'Börükçü iskeletlerinin paleoantropolojik incelenmesi' *19. Arkeometri Sonuçları Toplantısı, Milli Kütüphane Basımevi*, 27–40

Steele, D.G., McKern, T.W. 1969: 'A method for assessment of maximum long bone length and living stature from fragmentary long bone' *American Journal of Physical Anthropology* 3: 215–28

Stevenson, P.H. 1929: 'On racial differences in stature long bone regression formulae, with special reference to stature reconstruction formulae for the Chinese', *Biometrika*, 21.1–4: 303–21

Şam, B. 1994: *Fibula Uzunluğundan Vücut Boy Uzunluğunun Tahmini*. İstanbul, T.C. Adalet Bakanlığı, Adli Tıp Kurumu Başkanlığı, Uzmanlık Tezi

Şenyürek, M.S. 1949: 'Truva civarında Kumtepe'de bulunmuş iskeletlere dair bir not' *Ankara Üniversitesi Dil ve Tarih-Coğrafya Fakültesi Dergisi* 7.2: 295–304

Şenyürek, M.S. 1955: 'A note on the long bones of Chalcolithic Age from Yümüktepe' *Belleten* 74: 247–70

Telkka, A. 1950: 'On the prediction of human stature from the long bones' Acta Anatomica 9.1–2: 105–17

Telkka, A., Palkama, A., Virtama, P. 1962: 'Prediction of stature from radiographs of long bones in children' *Journal of Forensic Sciences* 7.4: 474–79

Trotter, M., Gleser, G.C. 1952: 'Estimation of stature from long bones of American whites and negroes' *American Journal of Physical Anthropology* 10: 463–514

— 1958: 'A re-evaluation of estimation of stature based on measurements of taken during life and of long bones after death' *American Journal of Physical Anthropology* 16: 79–123

Ubelaker, D.H. 1978: *Human Skeletal Remains*. Chicago, Smithsonian Institution, Aldine Publishing Company

Ulu, M.E.K., Öztürk, B., Atalay, K., Özkazanç, M.A., Terzi, Ö. 2020: 'Body height estimation based on magnetic resonance imaging measurements of the second cervical vertebra in the Turkish population' *Australian Journal of Forensic Sciences* DOI: 10.1080/00450618.2020.1745277

Vandervael, F. 1943: *Notions de biometrie humanie*. Liege, Desoer

Zorab, P.A., Harrison, A., Harrison, W.J. 1964: 'Estimation of height from tibia length' *The Lancet* 7633: 1063

4. Re-evaluation of the prevalence of porotic hyperostosis in Anatolian populations by applying the 'osteological paradox' approach

Handan Üstündağ

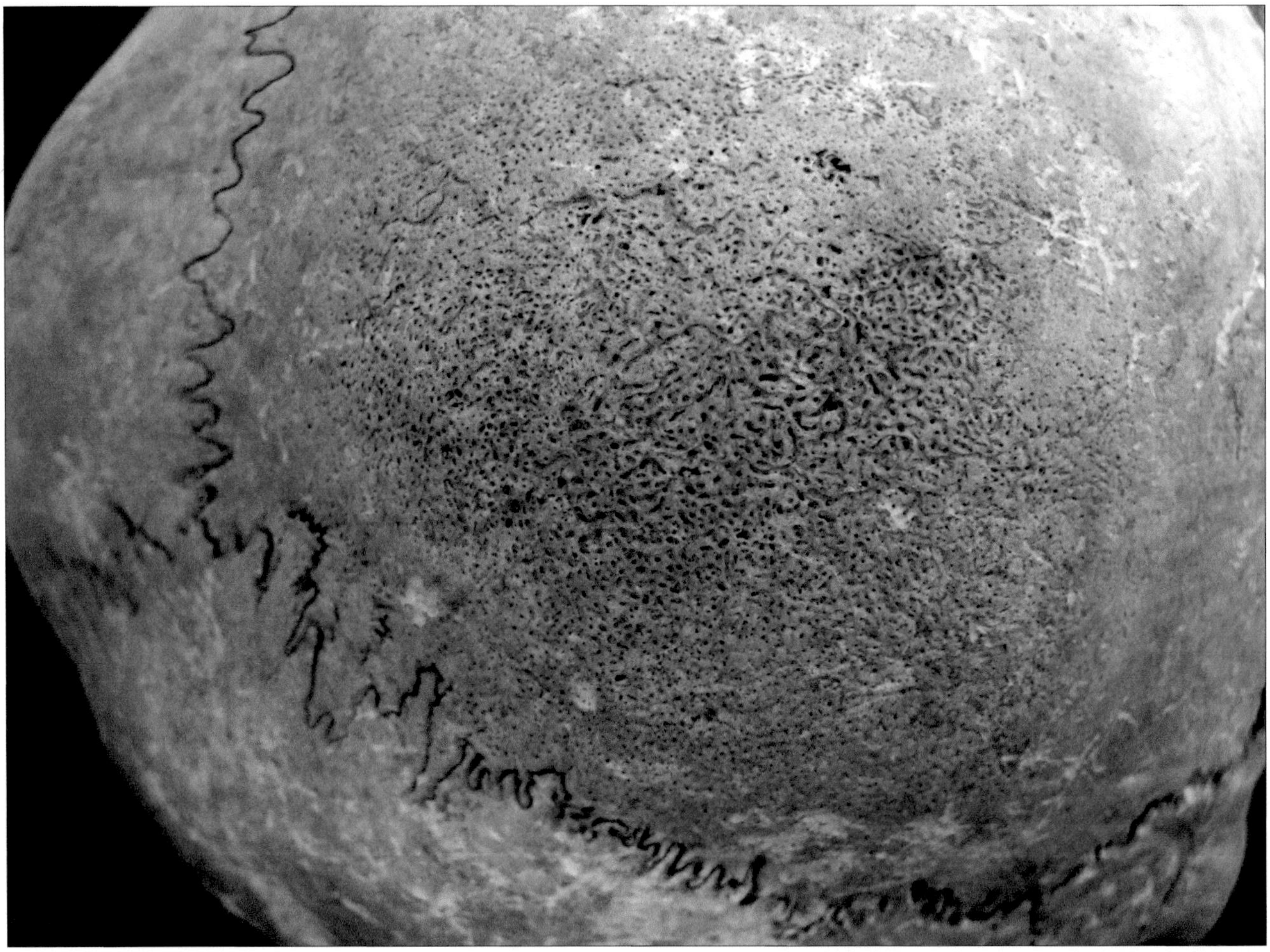

Fig. 4.1. Porotic hyperostosis (PH) on an individual from Kültepe. Kültepe human bones archive (photo by author).

Introduction

Porotic hyperostosis (PH), characterised by pitting and porosity in the cranial vault (Fig. 4.1), is one of the most frequently studied pathological conditions in bioarchaeology. It is widely accepted that childhood anaemia is the main cause of PH (Stuart-Macadam 1985, 1992; Blom et al. 2005; Walker et al. 2009; Brickley 2018). Porotic hyperostosis has also been frequently reported in Anatolian skeletal sample populations. Its prevalence in Anatolian populations, ranging from the Neolithic period to the post-medieval period, varies between 1% and 73% (Sevim 1998; Erdal 2000; Üstündağ 2011; Şahin 2019). Sevim (1998) suggested that ecological factors were responsible for these differences in prevalence, while Erdal (2011) suggested that this is due to inconsistency in data collection techniques. In addition to these, different demographic characteristics of the samples should also be mentioned as a factor in the large variation of PH prevalence rates.

There are only a few studies directly addressing PH or anaemia in Anatolian past populations (Sevim 1998; Üstündağ 2011; Büyükkarakaya, Erdal 2012; Özbek 2012). Porotic hyperostosis data in Anatolian populations are often presented in osteological or palaeopathological reports (e.g. Özbek 1989; Schultz 1989a; 1989b; Güleç et al. 1998; Erdal 2003; 2011; Erdal et al. 2003; Gözlük et al. 2003; Üstündağ, Demirel 2008; Üstündağ 2009a; 2009b; 2009c; 2014; 2019; Erdal, Özbek 2010; Özdemir, Sevim Erol 2010; Şahin 2019). Although some of these studies reported age-specific prevalence rates, these rates were not considered together with the demographic characteristics of the sample populations (e.g. Güleç et al. 1998; Erdal 2003, 2011; Erdal et al. 2003; Gözlük et al. 2003; Özdemir, Sevim Erol 2010; Üstündağ 2014). In addition, when comparing PH rates from various sample populations, the demographic characteristics of the relevant populations were generally not taken into account (e.g. Sevim 1998; Erdal 2000; Özdemir, Sevim Erol 2010; Üstündağ 2011; Şahin 2019). Only a few studies have attempted to correlate PH prevalence rates with demographic characteristics of the studied sample population (e.g. Üstündağ 2011; 2019; Şahin 2019). Presenting only subadult data (Schultz 1989b; Uysal-Uğur 1995; Erdal, Özbek 2010; Büyükkarakaya, Erdal 2012) or only adult data (Angel 1971) or just the crude prevalence (Üstündağ 2009b; 2009c), also makes it difficult to understand how the prevalence of PH in these populations varies between age groups.

The purpose of this study is to discuss why it is important to consider demographic characteristics (i.e. age distribution) of sample populations when comparing PH prevalence rates. Associating pathology with demography is not new to palaeopathology. An article discussing the 'osteological paradox' published nearly thirty years ago (Wood et al. 1992), and the subsequent debates, have dealt with this issue extensively (Goodman 1993; Cohen 1994). As a result, many palaeopathologists have begun to apply the osteological paradox approach to interpret pathological data in relation to demography (DeWitte, Stojanowski 2015). But for some reason this approach has never been applied by researchers in Türkiye. Therefore, this study attempts to demonstrate the benefits of using the osteological paradox approach when analysing PH data in a group of Anatolian samples.

The osteological paradox

The 'osteological paradox' is now a well-known approach in bioarchaeology that originally derived from the article by Wood and colleagues (1992). This article initiated a new debate on the interpretation of palaeopathological data, with an emphasis on heterogeneous frailty and selective mortality. The osteological paradox criticised the definition of 'healthy' in palaeopathology. It is confusing that the populations studied in palaeopathology consists of the dead, not the living. It would not be correct to assume that individuals whose bones do not show any pathological changes were healthy because they may have died before they could develop bone changes, even if they were not healthy. On the other hand, an individual with pathological changes on their bones could be considered as healthy because they lived long enough for changes to develop. Individuals differ in

Fig. 4.2. Map of Türkiye showing the locations of the sample populations (map by author).

their frailty, with some of those exposed to the same factors dying prematurely, while others survive, and individuals exhibiting changes on their bones may actually be healthier than individuals without any changes on their bones. This was seen by Wood and colleagues as a limitation of palaeopathological research. However, Goodman (1993) suggested that distinguishing those healthier survivors from non-survivors is possible by examining the prevalence of the pathology in relation to mortality. DeWitte and Stojanowski (2015) stated that many bioarchaeologists today should apply the osteological paradox approach to their research by considering subadults, demographic phenomena and the process of skeletal lesion formation.

Materials and methods

In this study, data derived from five Anatolian skeletal sample populations are examined: Aizanoi, Kültepe, Alanya Castle, Phokaia and Kadıkalesi (Fig. 4.2). The samples are dated to the Hellenistic and Roman and Byzantine periods. Porotic hyperostosis data of these sample populations were previously collected and published by the author (Üstündağ, Demirel 2008; Üstündağ 2009a; 2009b; 2009c; 2011; 2019). This study is an attempt to re-evaluate the existing data using a new approach.

Aizanoi

Aizanoi is located in Çavdarhisar county of Kütahya province, in inner-west Anatolia. It was an important city of the Phrygia region during the Roman period. There were a number of locations where skeletal remains were found at Aizanoi. In this study, only human skeletal remains from inhumations in the Northern Necropolis were included. The remains were unearthed between 2011 and 2014. The sample population consists of 80 individuals dating from the first century BC to the first century AD. Adults dominated the sample (Tab. 4.1), and the reason for this is that 25 individuals, mostly adults, were found together in a chamber grave (Üstündağ 2019).

Kültepe

Kültepe is located near the modern city of Kayseri, in Central Anatolia. The site, which was inhabited for a long time, is well known as the city of Kanesh in the Middle Bronze Age. The site consists of two main areas, the lower city and the mound. A large cemetery used in the Hellenistic and Roman periods (between the third century BC and sixth century AD) was found in the upper layer of the area of the lower city. During the 2007 and 2008 field seasons several burials were unearthed from this cemetery. The skeletons included in this study consist of 134 individuals from the 2007 excavation season (see Tab. 4.1 for the age distribution of the sample population) (Üstündağ 2009a).

Alanya Castle

Alanya Castle is located in the Alanya District of Antalya, on the Mediterranean coast of Anatolia. It has been inhabited since the Hellenistic period, but the ruins of the castle that can be reached today belong to the Seljuk period. Eight burials dating to the 10th century AD were unearthed in the central area of the inner castle. The overall skeletal sample population consists of 27 individuals. The sample population is dominated by adults and young adults (see Tab. 4.1) (Üstündağ, Demirel 2008).

Phokaia

Phokaia was an important Ionian city located in the Foça District of Izmir Province on the Aegean coast of western Anatolia. In 2007, a Byzantine cemetery was unearthed next to the Athena temple of Phokaia. The cemetery had two phases, with the graves of the first phase being dated to the 8th century AD, and those of the second phase being dated to the 10th–11th centuries AD (Özyiğit 2009). The skeletons examined in this study consist of 25 individuals belonging to the second phase of the necropolis. Seventeen (68%) of these individuals were adults, and most of those were young adults (Tab. 4.1) (Üstündağ 2009b).

Kadıkalesi

Kadıkalesi was a Byzantine castle located on the Aegean coast of western Anatolia, near the modern town of Kuşadası in Aydın. This castle was built in the 12th–13th centuries AD to protect the ancient city and harbour of Anaia (Mercangöz 2008). During the excavations carried out in Kadıkalesi between 2005 and 2007, skeletal remains were found around a church which had been built inside the city wall, and also on the outer edge of the city wall. The skeletal sample population consists of 58 individuals dating to the Late Byzantine period (AD 12–13th centuries). The Kadıkalesi sample population is dominated by subadults (60%). The proportion of infants, children and adolescents is significantly higher than the other sample populations examined in this study (Tab. 4.1) (Üstündağ 2009c; 2011).

Both subadult and adult skeletons were examined macroscopically for the presence or absence of PH, and PH was recorded when at least two-thirds of the cranial

	Aizanoi	*Kültepe*	*Alanya Castle*	*Phokaia*	*Kadıkalesi*
Foetal	0	1.5	3.7	4	3.4
Infant	5	12.7	7.44	8	17.2
Child	10	12.7	7.4	16	25.9
Adolescent	5	7.5	3.7	8	13.8
Young adult	17.5	20.9	44.4	40	24.1
Middle adult	15	17.9	7.4	12	3.4
Old adult	5	11.9	11.1	12	3.4
Adult	42.5	14.9	14.8	–	8.6
Total N	**80**	**134**	**27**	**25**	**58**

Tab. 4.1. Percentages (%) of individuals by age group in the sample populations. Foetal (<birth), infant (birth–2.9 years), child (3–11.9 years), adolescent (12–19.9 years), young adult (20–34.9 years), middle adult (35–49.9 years), old adult (50+ years), and adult (>19 years).

vault was present. The severity of the lesions was recorded according to the scale given in Stuart-Macadam (1985). The active and healed status of the lesions was also noted based on the criteria outlined in Mensforth et al. (1978). However, to enable comparability, all of these categories have been reduced to two categories, present and absent.

To compare age-specific prevalence rates, the sample populations were divided into two overall pooled age groups: subadults (birth–19 years) and adults (>19 years). Only these two large age categories were preferred because of the small sample sizes. Prevalence rates have been examined in subadults, adults, and subadults and adults combined. Then, the prevalence rates were evaluated together with subadult mortality rates.

In this study, the sex of the individuals was not taken into account since the main purpose was to determine how the age distribution of the sample populations affected the prevalence of PH.

Results

The prevalence rates of PH and subadult mortality rates (i.e. the proportion of subadults in the sample population) for the sample populations are presented in Table 4.2. When combining subadults and adults, the Aizanoi sample displays the highest prevalence of PH (55%), and the Kadıkalesi sample has the lowest (17.1%). The prevalence rates of PH in the combined group are quite similar for the Alanya Castle, Kültepe and Phokaia sample populations, and range between 30% and 36.4%.

For the Kadıkalesi sample population, the prevalence rate of PH was low in both subadults and adults (10% and 26.7%, respectively). In the Alanya Castle sample population, the prevalence rate of PH was higher in subadults (66.6%) than adults (27.3%). Conversely, the PH prevalence rate was higher in adults (38.5%) than subadults (14.3%) in the Phokaia sample population. The prevalence rates of PH were similar in subadults and adults in the Kültepe sample population (34.8% and 37%, respectively). In the Aizanoi sample population, it was not possible to calculate the prevalence rates of PH in subadults due to the low number of subadults and the poor preservation of their cranial bones. Therefore, the PH rate at Aizanoi is from only the adults.

The highest subadult mortality rate was found for the Kadıkalesi sample population (60%). Subadult mortality was low in both the Aizanoi and Alanya Castle sample populations (20% and 22%, respectively). The Kültepe and Phokaia sample populations had similar subadult mortality rates (34% and 36%, respectively), and this can be described as moderate compared to the other sample populations.

Discussion

Is PH an indicator of childhood anaemia?

The main suggested cause of PH is marrow expansion to aid red blood cell production that occurs during childhood anaemia (Stuart-Macadam 1992; Blom et al. 2005; Walker et al. 2009; Brickley 2018). Red blood cells are produced in red (hematopoietic) marrow. At birth all marrow is red, but fatty yellow marrow replaces it over time as the child grows up (Brickley 2018; O'Donnel et al. 2020). In adults, red blood cell production of red marrow occurs in the spongy bones of the vertebrae, sternum and ribs, while in children and adolescents it occurs in the diploe of the cranial vault bones, and in the medullary cavities of the long bones. Therefore, only childhood anaemia can lead to expansion of the diploe and the formation of PH on the cranial vault (Stuart-Macadam 1985; 1992; Blom et al. 2005; Walker et al. 2009; Brickley 2018). However, what

	PH prevalence						*Subadult mortality rate*
	Combined		*Subadults*		*Adults*		
	n/N	%	n/N	%	n/N	%	
Aizanoi	11/20	55	–	–	11/20	55	20
Kültepe	28/77	36.4	8/23	34.8	20/54	37	34
Alanya Castle	5/14	35.7	2/3	66.6	3/11	27.3	22
Phokaia	6/20	30	1/7	14.3	5/13	38.5	36
Kadıkalesi	6/35	17.1	2/20	10	4/14	26.7	60

Tab. 4.2. Prevalence of PH and subadult mortality rates in the sample populations.

type of anaemia was actually responsible for the high incidence of PH in past populations has always been a controversial issue. Angel (1966) suggested that PH in eastern Mediterranean skeletal samples is indicative of thalassemia, which provides protection against malaria. Palaeopathological studies in the 1970s and beyond consider iron deficiency anaemia to be the main cause of PH as a result of poor diet, gastrointestinal infections and parasitic infestations in children (Mensforth et al. 1978; Stuart-Macadam 1985; 1992; Ortner 2003). Later, Walker et al. (2009) challenged the iron deficiency hypothesis by arguing that iron deficiency inhibits bone marrow expansion rather than promoting it; he suggested that PH is indicative of hemolytic or megaloblastic anaemia. They argued that megaloblastic anaemia in children may be associated with maternal vitamin B12 and folate deficiencies or gastrointestinal infections at the time of weaning. In fact, the causal factors are basically the same in iron or vitamin B12 deficiency anaemia; insufficient animal protein in the diet or gastrointestinal parasites, or infections resulting from unsanitary living conditions. Therefore, some researchers prefer to define PH as an indicator of childhood nutrient deficiency (Mays 2018).

Hereditary anaemia such as thalassemia or sickle cell anaemia may also produce cranial porosity, although their incidence is relatively rare (Walker et al. 2009). Porotic lesions on cranial bones similar to PH can also occur due to infantile scurvy and rickets caused by vitamin C and vitamin D deficiencies, respectively. Both conditions can be differentiated from anaemia, as there are other types of changes in the cranial and post-cranial bones (Ortner, Ericsen 1997; Ortner, Mays 1998; Ortner 2003; Mays 2018). A link between PH and respiratory infection has also been proposed, but this appears to not be supported by further studies (O'Donnel et al. 2020).

Application of the osteological paradox approach to the interpretation of PH data

By using the osteological paradox approach, the following determinations can be made about the relationship between PH, anaemia and mortality:

1) PH is an indicator of childhood anaemia. It occurs in childhood, but remains later in life and can be observed in adults. Thus, both subadults and adults with PH lesions are considered to have experienced childhood anaemia. However, the presence of PH in subadult or adult skeletons should be interpreted differently.

2) Bone lesions take time to develop. Some children who were exposed to anaemia may have died before the bone lesions developed. In this case, whether an individual without PH can be assessed as healthy (i.e. not to have suffered from childhood anaemia) depends largely on the age of death. If the individual without PH is an adult, it is considered as healthy. However, if the individual without PH is a small child, it cannot be considered healthy because the child may have suffered anaemia but died before developing observable PH.

3) Individuals with PH (subadults or adults) were those who lived long enough to develop bone lesions (i.e. those who were healthier).

4) When interpreting the prevalence of PH, the subadult mortality rate should be taken into account. If the subadult mortality rate is high in a skeletal sample population, this indicates that many children may have died before they had time to develop bone lesions.

When we apply the traditional approach to compare the sample populations in the present study, we can state that anaemia was more common in the Aizanoi population (55% prevalence) but less common in the Kadıkalesi population (17%). However, it may be wrong to accept this simple conclusion without examining the age-specific prevalence and mortality rates of the sample populations.

In the Aizanoi sample population, there was no data on the prevalence of PH in subadults. Thus, the high prevalence of PH in the Aizanoi sample population is only indicative of its presence in adults. For this sample population, it would not be reliable to comment without knowing the prevalence rate of PH in the subadults.

In the case of Kadıkalesi, although the prevalence of PH is low, it would not be correct to suggest that childhood anaemia was rare in this population, as the subadult mortality rate was very high. It is possible that the children who were affected by anaemia died before developing any bone lesions. In fact, it is difficult for us to know exactly what the situation was. These children, however, can be described as frailer than other children who survived and lived long enough to develop bone lesions.

In the Alanya Castle sample population, although the prevalence rate of PH was higher in subadults than in adults, the subadult mortality rate was quite low. This suggests that many children suffered from anaemia, but there may be other reasons for the low subadult mortality rate in this small sample population.

The subadult mortality rates of the Kültepe and Phokaia sample populations are similar, and the PH prevalence rates are higher in adults than subadults in both sample populations. In this case, we can make a valid comparison between these two samples. The PH rate at Kültepe was higher than at Phokaia, especially in subadults. This suggests that childhood anaemia was probably more common in the Kültepe population than in the Phokaia population.

Conclusion

This study has revealed that age-specific prevalence and subadult mortality rates are useful for comparing populations. The prevalence rates in subadults and adults need to be evaluated separately because they indicate different conditions. A higher prevalence rate in adults indicates that several individuals suffered from anaemia in their childhood, but they survived into adulthood. On the other hand, a higher prevalence rate in subadults indicates that most children exposed to anaemia died prematurely. As shown by the sample populations examined in this study, an important factor influencing the prevalence rate of PH is the subadult mortality rate. If the PH prevalence rate is low but subadult mortality rate is high in a skeletal sample population, this indicates that many young children may have died before developing PH.

When we consider the age-specific prevalence and mortality rates in palaeopathological studies, we can reach more accurate conclusions and overcome the purported limitations of the osteological paradox.

Bibliography

Angel, J.L. 1966: 'Porotic hyperostosis, anemias, malarias, and marshes in the prehistoric Eastern Mediterranean' *Science* 153: 760–62

— 1971: 'Early Neolithic skeletons from Çatal Höyük: demography and pathology' *Anatolians Studies* 21: 77–99

Blom, D.E., Buikstra, J.E., Keng, L., Tomczak, P.D., Shoreman, E., Stevens-Tuttle, D. 2005: 'Anemia and childhood mortality: latitudinal patterning along the coast of pre-Columbian Peru' *American Journal of Physical Anthropology* 127.2: 152–69

Brickley, M.B. 2018: 'Cribra orbitalia and porotic hyperostosis: a biological approach to diagnosis' *American Journal of Physical Anthropology* 167: 896–902

Büyükkarakaya, A.M., Erdal, Y.S. 2012: 'Eski Anadolu toplumlarında anemi' in E. Güleç, İ. Özer, M. Sağır, B. Koca Özer (eds), *Biyolojik Antropoloji, Ankara Üniversitesi Dil ve Tarih-Coğrafya Fakültesi 75. Kuruluş Yıldönümü Anı Kitabı*. Ankara, Ankara Üniversitesi Dil ve Tarih-Coğrafya Fakültesi Yayınları: 137–62

Cohen, M.N. 1994: 'The osteological paradox reconsidered' *Current Anthropology* 35: 629–31

DeWitte, S.N., Stojanowski, C.M. 2015: 'The osteological paradox 20 years later: past perspectives, future directions' *Journal of Archaeological Research* 23: 397–450

Erdal, Y.S. 2000: 'Eski Anadolu toplumlarında çocuk sağlığı ve hastalıkları' *Çocuk Sağlığı ve Hastalıkları Dergisi* 43: 5–19

— 2003: 'Büyüksaray-Eski Cezaevi çevresi kazılarında gün ışığına çıkarılan insan iskelet kalıntılarının antropolojik analizi' *Arkeometri Sonuçları Toplantısı* 18: 15–30

— 2011. 'Tasmasor Yakınçağ Nekropolü ve iskeletlerinin antropolojik analizi' in S.Y. Şenyurt (ed.), *Tasmasor*. Ankara, Bilgin Yayınları: 329–458

Erdal, Ö.D., Özbek, M. 2010: 'Değirmentepe (Malatya) çocuk iskeletlerinin antropolojik analizi' *Arkeometri Sonuçları Toplantısı* 25: 279–96

Erdal, Ö.D., Eroğlu, S., Erdal, Y.S., Büyükkarakaya, A. 2003: 'Şaşal/İzmir iskelet topluluğunun paleopatolojik ve demografik analizi' *Arkeometri Sonuçları Toplantısı* 18: 1–14

Goodman, A. 1993: 'On the interpretation of health from skeletal remains' *Current Anthropology* 34: 281–88

Gözlük, P., Yılmaz, H., Yiğit, A., Açıkkol, A., Sevim, A. 2003: 'Hakkari Erken Demir Çağı iskeletlerinin paleoantropolojik açıdan incelenmesi' *Arkeometri Sonuçları Toplantısı* 18: 31–40

Güleç, E., Sevim, A., Özer, İ., Sağır, M. 1998: 'Klazomenai'de yaşamış insanların sağlık sorunları' *Arkeometri Sonuçları Toplantısı* 13: 133–59

Mays, S. 2018: 'Micronutrient deficiency diseases: anemia, scurvy, and rickets' in W. Trevathan (ed.), *The International Encyclopedia of Biological Anthropology* Vol. 1. Hoboken, Wiley-Blackwell: 1–5

Mensforth, R.P., Lovejoy, C.O., Lallo, J.W., Armelagos, G.J. 1978: 'The role of constitutional factors, diet, and infectious disease in the etiology of porotic hyperostosis and periosteal reactions in prehistoric infants and children' *Medical Anthropology* 2.1: 1–57

Mercangöz, Z. 2008: 'Kuşadası, Kadıkalesi Kazısı 2006 yılı çalışmaları' *Kazı Sonuçları Toplantısı 29.3*: 449–70

O'Donnel, L., Hill, E.C., Anderson, A.S.A., Edgar, H.J.H. 2020: 'Cribra orbitalia and porotic hyperostosis are associated with respiratory infections in a contemporary mortality sample from New Mexico' *American Journal of Physical Anthropology* 173: 721–33

Ortner, D.J. 2003: *Identification of Pathological Conditions in Human Skeletal Remains*. San Diego, Academic Press

Ortner, D.J., Ericksen, M.F. 1997: 'Bone changes in the human skull probably resulting from scurvy in infancy and childhood' *International Journal of Osteoarchaeology* 7: 212–20

Ortner, D.J., Mays, S. 1998: 'Dry-bone manifestations of rickets in infancy and early childhood' *International Journal of Osteoarchaeology* 8: 45–55

Özbek, M. 1989: 'Çayönü insanları ve sağlık sorunları' *Arkeometri Sonuçları Toplantısı* 4: 121–52

— 2012: 'Kronik demir eksikliğine bağlı anemi: klinik ve paleopatolojik bulgular ışığında yeni değerlendirmeler' in A.A. Akyol, K. Özdemir (eds), *Türkiye'de Arkeometrinin Ulu Çınarları Prof. Dr. Ay Melek Özer ve Prof. Dr. Şahinde Demirci'ye Armağan*. İstanbul, Homer Kitabevi ve Yayıncılık: 269–80

Özdemir, S., Sevim Erol, A. 2010: 'Minnetpınarı iskeletlerinin paleopatolojik açıdan analizi' *Antropoloji* 23: 95–127

Özyiğit, Ö. 2009: '2007 Yılı Phokaia kazı çalışmaları' *Kazı Sonuçları Toplantısı* 30.1: 390–410

Şahin, S. 2019: 'Dilkaya (Orta Çağ) insanlarının sağlık yapısı' *Antropoloji* 37: 50–71

Schultz, M. 1989a: 'Osteologische Untersuchungen an den spätmittelalterlichen Skeletten von Pergamon – ein vorlaufiger Bericht' *Arkeometri Sonuçları Toplantısı* 4: 111–14

— 1989b: 'Der Gesundheitszustand der frühbronzezeitlichen Bevölkerung vom İkiztepe. 1. Kinderskelette' *Arkeometri Sonuçları Toplantısı* 4: 115–18

Sevim, A. 1998: 'Eski Anadolu toplumlarında gözlenen bir patolojik doku bozukluğu: porotic hyperostosis' *Antropoloji* 13: 229–44

Stuart-Macadam, P. 1985: 'Porotic hyperostosis: representative of a childhood condition' *American Journal of Physical Anthropology* 66: 391–98

— 1992: 'Porotic hyperostosis: A new perspective' *American Journal of Physical Anthropology* 87: 39–47

Üstündağ, H. 2009a: 'Kültepe/Kanesh (Turkey), season 2007' *Bioarchaeology of the Near East* 3. Short Fieldwork Reports: 31–35

— 2009b: 'Phokaia (Turkey), season 2007' *Bioarchaeology of the Near East* 3. Short Fieldwork Reports: 27–31

— 2009c: 'Kuşadası, Kadıkalesi/Anaia'da bulunan insan iskelet kalıntıları' *Arkeometri Sonuçları Toplantısı* 24: 209–28

— 2011: 'Genetik, çevresel ve kültürel etmenlerin ışığında anemi: bazı eski Anadolu topluluklarından örnekler' *TUBA-AR* 14: 173–92

— 2014: 'Human remains from Kültepe-Kanesh: preliminary results of the Old Assyrian burials from the 2005–2008 excavations' in L. Atici, F. Kulakoğlu, G. Barjamovic, A. Fairbairn (eds), *Current Research at Kültepe-Kanesh. An Interdisciplinary and Integrative Approach to Trade Networks, Internationalism, and Identity*. Atlanta, Lockwood Press: 157–76

— 2019: 'Aizanoi kazılarında bulunan insan iskelet kalıntıları / Human skeleton remains found in Aizanoi excavations' in E. Özer (ed.), *Aizanoi IV. Anadolu'da Hellenistik ve Roma Dönemlerinde Ölü Gömme Adetleri Uluslararası Sempozyumu Bildiri Kitabı (23–26 Temmuz 2018) / The Book of International Symposium on Burial Customs in Anatolia During the Hellenistic and Roman Periods (July 23–26, 2018)*. Ankara, Bilgin Kültür Sanat Yayınları: 311–29

Üstündağ, H., Demirel, F.A. 2008: 'Alanya Kalesi kazılarında bulunan insan iskelet kalıntılarının osteolojik analizi' *Arkeoloji ve Etnoğrafya Dergisi* 8: 79–90

Uysal Uğur, G. 1995: 'Oylum Höyük çocuklarının paleopatolojik açıdan analizi' *Hacettepe Üniversitesi Edebiyat Fakültesi Dergisi* 12.1–2: 187–206

Walker, P.L., Bathurst, R.R., Richman, R., Gjerdrum, T., Andrushko, V.A. 2009: 'The causes of porotic hyperostosis and cribra orbitalia: a reappraisal of the iron-deficiency-anemia hypothesis' *American Journal of Physical Anthropology* 139.2: 109–25

Wood, J.W., Milner, G.R., Harpending H.C., Weiss, K.M. 1992: 'The osteological paradox: problems of inferring prehistoric health from skeletal samples' *Current Anthropology* 33.4: 343–70

5. Biological distance relations in Anatolian populations: an anthropological review

Serpil Eroğlu

Introduction

The investigation of biological relations between human groups, which has long been one of the most important fields of interest in anthropology, provides significant information about a population's history and its relations with others based on the morphology and size of bones and teeth unearthed from archaeological excavations. Biological distance studies, which aim to establish genetic differences between populations by using both genetic and environmentally-affected polygenic traits (Buikstra et al. 1990; Pilloud, Hefner 2016), generally focus on two main areas. The first of these is population structure, which implies gene flow between populations, genetic drift, migration, founder effects, and the second is the distribution and size of a population, namely population history, which focuses on the degree of similarity between gene flow and populations from a common ancestor (Relethford 1996). This approach takes into account the models of micro-evolution theory in order to reconstruct population history and population structure (Knudson, Stojanowski 2008). Biological distance analysis identifies patterns of statistical variation within and between populations through the simultaneous evaluation of multiple features using several multivariate techniques (Pilloud, Hefner 2016; Nikita 2017). Such analysis is not only about who is related to whom, but also how these relationships have changed over time, and the potential importance of socially increasing or decreasing biological integration (Stojanowski, Schillaci 2006; Pilloud, Larsen 2011).

Phenotypic data in skeletons generally include metric and non-metric traits of the cranium and/or teeth. They are morphological variants whose expression cannot be measured in a linear manner (Pilloud, Hefner 2016). Although ancient DNA (aDNA) studies have yielded reliable results by overcoming the early problems regarding the organic and inorganic contamination of the samples (Williams et al. 2002; O'Rourke, Enk 2012), isolation issues related to the conservation status of the skeletons are still encountered. Nevertheless, despite the rapid development of aDNA and stable isotope studies on past mobility in the field of bioarchaeology, today, biological distance studies still maintain a relevant role. Macroscopic methods can be applied regardless of the degree of diagenesis and contamination of skeletal and dental tissues, and the inclusion of both skull and postcranial bones in the analyses, as well as the ease of data collection and the low cost of the analysis, are the main reasons why skeletons and teeth are still preferred in biological distance studies (Nikita 2020). Moreover, inheritance studies on humans (Susanne 1975; 1976; Sjovold 1984) and animals (Self, Leamy 1978; Cheverud, Buikstra 1981; 1982; Carslon 2006a; 2006b) have shown that cranial metric and non-metric traits are able to reflect the genetic structure to a certain extent.

There has been a long-lasting discussion that skeletal morphological traits cannot be used in biological distance studies due to environmental effects (nature over nurture) (Eroğlu 2011). However, according to Scott and Turner (1997), despite the fact that environmental factors may affect individual phenotypic expression to some extent, they do not have a significant effect on the frequency of a trait at a population level. Thus, aDNA studies, widespread from the 2000s, seem to have put an end to this debate in a way (Ricaut et al. 2010; Ottoni et al. 2011; Herrera et al. 2014; Hubbard et al. 2015).

It is known that studies on the classification of humans via the skull began during the Enlightenment in the 18th century (Brace 2010; Molnar 2015). In the same period the concept of 'race' was formalised, and various 'racial classification systems' were proposed. Defined as a typological characterisation of human variation, 'race' became a dominant topic in physical anthropology until the mid-20th century (Brace 2010; Little, Sussman 2010). However, with the establishment of 'New Physical Anthropology' in the 1950s, new research questions arose (Stini 2010). Rather than biological distance analysis, mass migrations and typological racial determinations, past population structure and short-range mobility began to be examined (Buikstra 1979; Little, Sussman 2010; Stini 2010). These developments were also accompanied by the development of statistical methods in the evaluation of phenotypic data (Hefner et al. 2016). Today, bioarchaeological and bioanthropological data are used to obtain information about lifestyle and relations between populations that lived in the past. The concept of bioarchaeology first entered the literature in 1972 and was later, in terms of the definition of the field used in the United States, defined to refer to the study of human remains from archaeological contexts, and began to be used for population studies based on human remains in 1977 (Buikstra 1977). Stating that bioarchaeological studies should be population-oriented, Larsen (2002) emphasised the necessity of an interdisciplinary field of study, where biological anthropology and archaeology work together.

The biological distance studies which have been conducted in Anatolia are generally not very well known, and are still often excluded from literature reviews of the field. For example, in her review study, 'Biodistance Studies in the Eastern Mediterranean and Middle East: An Overview and Future Prospects', Nikita (2020) mentions only a few biological distance studies related to Anatolia and conducted on Anatolian populations. Therefore, the present study aims to introduce biological distance studies conducted in Anatolia, from the establishment of biological anthropology as a field in Türkiye to the present day.

Biological distance studies in Anatolia

Studies before 1980

With the description of physical features of Turkish populations as one of the main goals, as well as all over the world, physical anthropology in Türkiye (Özbek 2001) began with studies into 'race'. From the commencement of physical anthropology studies in Türkiye (1925) until approximately 1980, with the exception of a few studies aimed at establishing relations between populations and at questioning the significance of differences, as well as of characteristic morphological traits (Krogman 1937; Şenyürek 1941; 1951; 1952; Cappieri 1969; 1970; 1972; Clement et al. 1974), studies on the phenotypic structure of Anatolians based on cranial and postcranial measurements have generally been typological, and skeletons unearthed from various settlements have been classified according to cephalic indices.

In Anatolia, three basic phenotypes were anthropologically identified between 1925 and 1980: robust and gracile Mediterranean, Alpine and Dinaric (Özbek 1990). The results of those early investigations revealed that the first human phenotype in Anatolia was the robust Mediterranean type, while the gracile Mediterranean and the Alpine types emerged through autochthonous development (Cappieri 1970). In the Neolithic period, especially in the skeletal populations of Çayönü (Özbek 1990) and Aşıklı (Özbek 1992; 1993), robust and gracile Mediterranean were the dominant phenotypes, while an increase in the ratio of the Alpine group representativeness was observed in the skeletal population of Çatalhöyük (Ferembach 1972). The Dinaric phenotype was observed in the Bronze Age skeletal populations from Karataş (Mellink, Angel 1968), Bodrum-Müsgebi (Çiner 1964) and Kusura (Kansu, Atasayan 1939).

The origins and causes of the process of a rounding of the skull (brachycephalisation) – which is a characteristic trait of the Alpine group, as well as of the gracile Mediterranean one – has also been long discussed (Cappieri 1969; 1970; 1972; Özbek 1990). Only 1 in 18 individuals of the Çayönü population, which included both robust and gracile Mediterranean individuals, dated to the Neolithic period, and it was found to be brachycephalic. A considerable number of brachycephalic individuals (22.9%) were found at the Neolithic site of Çatalhöyük, with a later date when compared to those from Çayönü (Clement et al. 1974). According to these studies, the morphology of the skull of Anatolians changed over time. In fact, craniofacial morphology, which was relatively homogenous in the Neolithic period, acquired a more heterogeneous characterisation with the advent of the Bronze Age (Şenyürek 1941; 1952; Krogram 1949; Cappieri 1969; 1970; 1972; Clement et al. 1974; Özbek 1990). Two main hypotheses were put forward to explain this issue: microevolution (Cappieri 1970; 1972; Clement et al. 1972; Özbek 1990) and migration (Krogram 1937; 1949; Şenyürek 1941; 1951; 1952; Mellink, Angel 1968).

The fact that the ratio of brachycephalic individuals gradually increased over time, starting from the Neolithic, led to a questioning of the reasons behind the emergence of the Alpine group in Anatolia, characterised by a brachycephalic head shape. It was claimed that the Alpine group emerged through a microevolutionary process rather than as a result of migration (Ferembach 1972; Clement et al. 1974; Özbek 1990). Clement et al. (1974) stated that 47 individuals from the Çatalhöyük population were of the robust and gracile Mediterranean phenotypes. The significant representation of the Alpine phenotype suggested that this group was a side branch which developed in the Mediterranean gene pool. Cappieri (1970; 1972), who studied the phenotypic diversity between and within groups of Anatolian and non-Anatolian populations living between 4000 and 1500 BC, found that Anatolian populations were largely homogenous and had a similar genotype. Cappieri (1969), who also conducted a biometric study on seven Asian regional types (Troad, Anatolian, Syro-Palestinian, Mesopotamian, Turkmen and Indian), stated that all the evidence and the anthropometric data indicated that a branch of the Mediterranean group lived in West Asia within a range stretching from the Aegean Sea to Bengal, and from the Caspian Sea to the Krishna River in Deccan during the late-Neolithic and pre-Iron Age. Individuals of the Mediterranean type were the only habitants of this vast region.

On the other hand, researchers who claimed that changes in the structure of Anatolian populations were due to migration (Krogram 1937: 49; Şenyürek 1941; 1951; Mellink, Angel 1968) investigated the origins of the different groups present in Anatolia. Şenyürek (1941), based on an examination of the skulls belonging to the Alişar, Kusura, Arslantepe and Hisarlık III populations dated to the Hittite period, stated that during the Chalcolithic period and Bronze Age the Anatolian population was mostly composed of dolichocephalic and mesocephalic individuals, while the rate of brachycephalic individuals was 16%. Therefore, during these periods, no significant changes seem to have occurred in the genetic structures of Anatolian populations. Nevertheless, the fact that the rate of brachycephaly increased to 42% in the Bronze Age may imply that it originated with individuals of the Alpine phenotype that migrated to Anatolia. He claimed that the Alpine phenotype invaders, who arrived in Anatolia in 2000 BC, were the Hittites (Şenyürek 1941). Şenyürek (1949) also expressed a similar opinion for Kumtepe, stating that the rate of brachycephaly increased in Anatolia during the Chalcolithic period via migration. Şenyürek (1952) stated that at Kültepe II, dated to the Bronze Age, and at Çatalhöyük and Tell Al-Judaiah, males were dolichocephalic and females were brachycephalic. According to Şenyürek (1952), the craniological evidence showed that the Hittites had already arrived in Anatolia when the Assyrian trade colony developed at Kültepe. Krogman (1949) suggested that some men belonging to the Çatalhöyük and Tell Al-Judaiah populations moved from south of the Taurus Mountains to the north between 1600 and 1200 BC and married Hittite women. Moreover, he stated that the people of Alişar Höyük were long-headed before the Early Bronze Age, with increasing heterogeneity being expressed in more round-headed individuals. According to Şenyürek (1951), since dolichocephalic and mesocephalic individuals found at Alacahöyük were similar to those from other Anatolian settlements, it may be suggested that they belonged to the native Bronze Age population of Anatolia. According to the same scholar (Şenyürek 1951), the population of this settlement was most probably composed of aristocrats and invaders. Mellink and Angel (1968), who studied the morphology of the Karataş skeletal population dated to the Early Bronze Age, reached a similar conclusion. In fact, the researchers found that although dolichocephaly was predominant, as at Çatalhöyük, the rates of gracile Mediterranean, Alpine and Dinaric groups were high as well. In addition, they identified a Nordic-Iranian influence, and stated that the Karataş population differed from the Macedonian and late Aegean ones, from which it originated, while correlating with populations from the east, such as Cypriots.

Krogman (1949), examined 31 skulls from Amik settlements (Çatalhöyük and Tell Al-Judaiah in Hatay) and identified a heterogenous group with both round-headed (Alpine and Armenoid phenotype) and long-headed (European and Mediterranean phenotypes) individuals, the latter type being predominant. He suggested that the Mediterranean phenotype (before 4000 BC) was in the region long before the Europeans (2500 BC), the Alpines (1600 BC) and possibly the Armenoids (500 BC), who originated from the Aral-Caspian region. He suggested that the Armenoid phenotype (short head), found in Anatolia as early as 1500 BC, developed from the broad head (Alpine) phenotype through genetic selection. Krogman also suggested, comparing the Early Bronze Age and Hittite skeletal populations of Alişar, Alacahöyük, Tepe Hisar and Sialk, the Alpine phenotype was representative of a large group of round-headed people, who was definitely not native, having entered Asia Minor later.

The third morphological group, the Dinaric phenotype (Krogman 1937; 1949; Cappieri 1970; 1972; Özbek 1990), characterised by a brachycephalic head shape like the Alpines, became part of the Anatolian population structure after the Alpines, and began to be observed from the Bronze Age onwards (Özbek 1990). It was stated that this group might have migrated into Anatolia (Krogman 1937; Cappieri 1972; 1979; Özbek 1990). The migration hypothesis is generally supported by more recent studies as well (Açıkkol et al. 2005; Eroğlu, Erdal 2006; Pilloud 2009; Seguchi et al. 2010; Philbin, Pilloud 2018; Adams 2020), which have suggested that migrations to Anatolia were quite limited, especially in the Neolithic and Chalcolithic periods, while a rapid migration wave characterised the Bronze Age, during which the Anatolian genetic structure became a heterogeneous one.

In general, although typological studies conducted before the 1980s had different research questions compared to today's biological distance perspective, the data obtained from them can still be evaluated using a modern approach.

Studies after 1980

In the 1980s, only three studies could be categorised as being biologically distance related. The first two were based on non-metric (Klung, Wittwer-Backofen 1983) and metric (Wittwer- Backofen 1987) analysis of the

Lidar Höyük skeletons, while the other one was based on the cranial metric traits of the İkiztepe skeletons (Wittwer-Backofen 1985). In the first study, Klung and Wittwer-Backofen (1983), by applying MMD analysis on 67 non-metric traits of 10 sample groups from Europe and the Near East, including Lidar Höyük, tried to determine differences between populations. In the second one, Wittwer-Backofen (1987) researched intra-population variations on skeletons dated to the Early and Middle Bronze Age and Roman periods of Lidar Höyük using Penrose's cluster analysis. In addition, Neolithic skeletons from Nevali Çori were used as a comparison group. According to his results (Wittwer-Backofen 1987), the Neolithic group clustered separately while similarities existed between different cultural layers in Lidar.

Wittwer-Bacofen's study (1985) represents the most detailed investigation in terms of biological distance, and we can now assess it further by including archaeological (Bilgi 2001), anthropological (Eroğlu 2005; Eroğlu, Erdal 2006) and ancient DNA (Skourtanioti et al. 2020) data. For the origin of the İkiztepe population during the Late Chalcolithic period (Welton 2010) two main views exist: one asserting that the İkiztepe population were natives of Anatolia (Bilgi 2001), and the other suggesting that the İkiztepe population came to the central Black Sea region through migration (Wittwer-Backofen 1985). Based on archaeological findings, Bilgi (1984; 1990) stated that before 2100 BC, 90% of the metal findings followed a local development, and that different cultural layers are characterised by continuity (Bilgi 2001). This observation suggested that the population of the region did not migrate from somewhere else. According to Bilgi (1990; 2001), since cultural artefacts such as potters' wheels and shaft-hole axes – which originated from northern Syria and northern Mesopotamia – were found at İkiztepe, even if a migration occurred, its origin should be from the south rather than from the Caucasus or the northern Black Sea region. In addition, according to the similarity of the pottery found at settlements such as Alişar, Acemhöyük, Karahöyük and İkiztepe, Bilgi argues that Middle Black Sea populations, including the İkiztepe population, originated from a Proto-Indo-European autochthonous tribe that lived in the region for thousands of years.

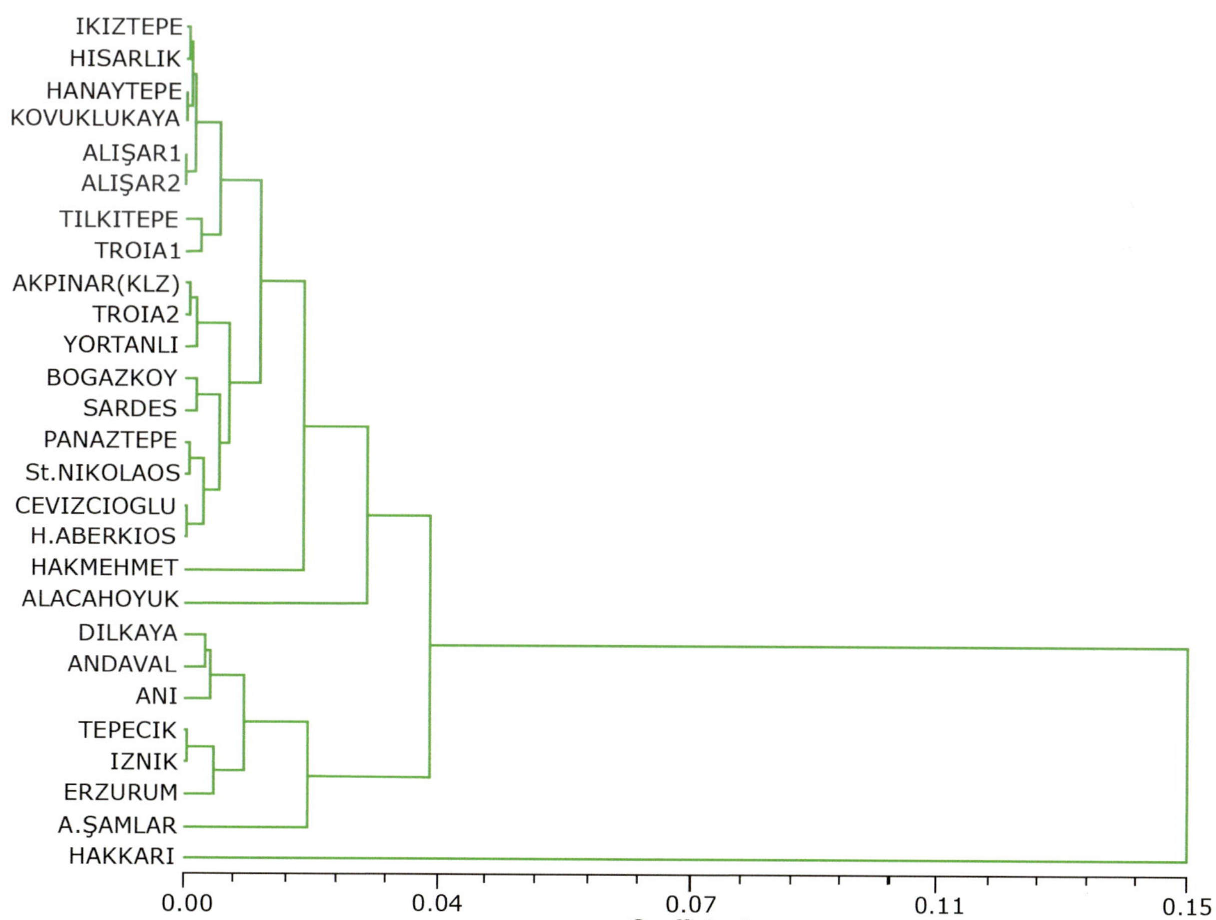

Fig. 5.1. Fenogram showing the hierarchical clustering of Anatolian groups (adapted from Eroğlu, Erdal 2006).

Wittwer-Backofen (1985), who claimed that the İkiztepe people migrated to that region, compared its population with those of the Balkans and the Near East by using Penrose's shape and size analyses. Based on the cranial measurements of the skeletons excavated from the necropolis of İkiztepe, he clustered the İkiztepe population together with those in southeast Europe, the coast of the Black Sea in southern Russia, at Cernavoda (Romania) and Vinca (Bulgaria). According to Wittwer-Backofen (1985) similarities in the use of ochre on the skeletons from southern Russia and those found at İkiztepe may support this view.

Eroğlu and Erdal (2006) conducted a biological distance analysis on the İkiztepe population and on other Bronze Age populations living in and around Anatolia by analysing six craniofacial metric traits. According to the results of this study (see Figs 5.1 and 5.2), the İkiztepe population is clustered together with the western Anatolia (Troad) populations, and with a Syrian (Ras Shamra) population, as well as with some other Mesopotamian populations. Thus, the relationship with the populations of the Balkans, the northern Black Sea region and the Caucasus was not identified due to the lack of a direct comparison. Conversely, a relationship was recognised between the regions to the south of Anatolia. Twenty-one craniofacial measurements belonging to Anatolian populations were divided into four groups, including İkiztepe as Late Chalcolithic, as well as those from the Byzantine, Ottoman and Modern periods, and these were compared with data from 37 populations living all over the world (Seguchi et al. 2010). It was found that the İkiztepe and other Anatolian populations were related to those from West Asia, Europe and the Caucasus. Moreover, it was observed that women constituted a more homogenous group, while men exhibited a more heterogeneous genetic makeup, being clustered together with Eurasian and Near Eastern populations (Seguchi et al. 2010).

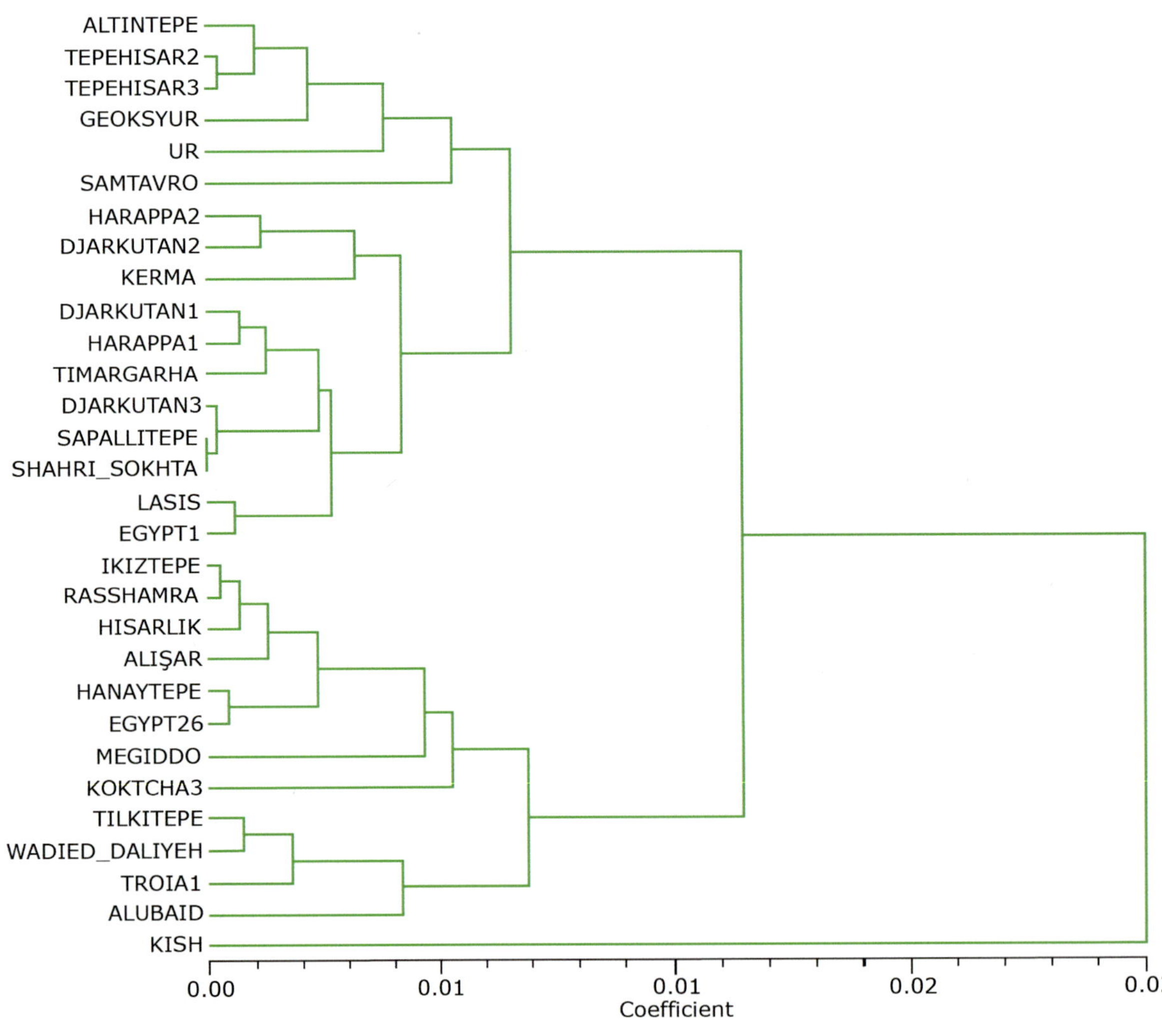

Fig. 5.2. Fenogram showing hierarchical clustering of Anatolian and adjacent groups (adapted from Eroğlu, Erdal 2006).

In an aDNA study of 110 individuals from Anatolia, the northern Levant and southern Caucasus dated from the Late Neolithic to the Late Bronze Age, including İkiztepe (Skourtanioti et al. 2020), it was concluded that a genetic continuity existed in northern, central and eastern Anatolia during those chronological periods, and that this situation is supported by the genetic similarity of the later populations, as well as the absence of new lineages after the Neolithic period. In addition, the cultural similarities observed in the pottery and metal findings between northern and central Anatolia and the Balkans – especially the Cernavoda region in Romania that seem to have continued until 3000 BC, as claimed by Thissen (1993) – were not supported by a gene flow. Thus, the results of the investigation conducted by Skourtanioti et al. (2020) contradicted the previous hypothesis, suggesting that the İkiztepe region had genetic relations with the Balkans, as well as intense cultural interaction with the populations living all around the Black Sea. In fact, this conclusion reflects the hypothesis of Bilgi (2001), who asserted that there was no direct external influence before 2100 BC, and that cultural products were local. It is worth mentioning that a comparison of the quantity and quality of the grave goods found in the necropolises of İkiztepe, Varna and Durankulak has revealed many differences, with the exception of bracelet- and ring-like objects (Bilgi 2003). Despite all of these studies, a more comprehensive investigation in which both cranial and dental data from the Balkans, the northern Black Sea region, the Caucasus and the Middle East are evaluated together is necessary in order to more comprehensively outline the genetic structure and history of the İkiztepe population.

While biological distance studies were not so common in the 1990s, a new phase in this field of research started in the 2000s (Özer, Güleç, 2000; Açıkkol et al. 2005; Eroğlu, Erdal 2006; Ricault, Waelkens 2008; Pilloud 2009; Seguchi et al. 2010; Eroğlu 2011; Pilloud, Larsen 2011; Eroğlu 2016; 2019; Adams 2020; Akbaba 2020), together with an increase in aDNA studies (Mathieson 2015; Kılınç et al. 2016; Feldman et al. 2019; Skourtanioti et al. 2020) in more recent years.

One of the first studies of the 2000s was carried out by Özer and Güleç (2000). Based on metric traits, they compared Anatolian populations dated between the Neolithic and Middle Ages with the medieval skeletons of Dilkaya. They concluded that the Dilkaya population was similar to the other populations dated to the same chronological period. A similar study was conducted by Açıkkol et al. in 2005. They compared the metric traits of ancient Anatolian populations dated from the Neolithic period to the Middle Ages with those of European, Asian and Near Eastern populations. They suggested that Bronze Age populations had a heterogeneous structure and were heavily influenced by gene flow from the Middle East and the Caucasus, while the Neolithic ones had a more homogenous genetic make-up. Also, post-Bronze Age populations were also considered to be heterogeneously structured. Biological distance analysis based on metric and non-metric traits of skeletons belonging to 12 populations (Late Chalcolithic to 20th century) coming from different regions of Anatolia (Eroğlu 2011), has also supported the suggestion of a heterogeneous genetic structure of mediaeval Anatolian populations, which clustered away from those of İkiztepe and Cevizcioğlu Çiftliği that are dated to earlier chronological periods.

Biological distance studies based on the tooth morphology of Anatolian skeletal populations (Eroğlu, Erdal 2009; Pilloud 2009; Pilloud, Larsen 2011; Pilloud et al. 2017; Akbaba 2020) began to contribute to the anthropological literature much later than studies based on cranial metric and non-metric traits. The first of these studies in Anatolia was Pilloud's doctoral thesis (2009) on the skeletal population from Çatalhöyük. Pilloud (2009), by analysing both metric and non-metric dental traits, carried out a biological distance analysis study with the aim of reconstructing and understanding the social organisation of the settlement. It was found that the biological affinity of the individuals within the households was very low. It was suggested that unrelated people from all over the settlement gathered in houses within the framework of social principles, and that kinship crossed biological boundaries and created a wider definition of it. In the same study, she compared the data from Neolithic Çatalhöyük with that from Aşıklı Höyük, Musular and later-period (AD 60–1650) Çatalhöyük. While, at first, the three Neolithic populations demonstrated low variation, most probably resulting from a gene flow into central Anatolia, the similarity of the later series to the Neolithic populations was significant. Pilloud and Larsen (2011), investigating burial kinship patterns, analysed the deciduous and permanent teeth of 266 individuals from Çatalhöyük. They concluded that individuals of a single household were not biologically related; in other words, their kinship was not biological, and instead was based on social and cultural circumstances. They interpreted this result as a sign of a practical kinship model due to the large workload based on plant production and animal husbandry that required a large number of people. In another study at Çatalhöyük (Larsen et al. 2019), adult male and female diets were found to be largely similar. Nevertheless, individuals buried in the same houses showed isotopically different diets, which provide corroborating results consistent with that of Pilloud and Larsen (2011). Moreover, the results of dental phenotypic analysis of the Çatalhöyük, Aşıklı Höyük and Musular populations were generally similar, indicating a patrilocal post-marital settlement practice (Pilloud et al. 2017). Philbin and Pilloud (2018) evaluated dental morphological data from four settlements dated to the Neolithic period (Boncuklu, Aşıklı, Musular and Çatalhöyük), and also from the later periods of occupation at Çatalhöyük. They concluded that the populations had a local development, and that gene flow increased over time.

In a recent study, Adams (2020) analysed dental morphological data of the Karataş-Semayük skeletons dated to the EBA to examine biological diversity in the region, the relationship of Karataş with other settlements in the Eastern Mediterranean, as well as kinship patterns. It was found that a significant biological diversity existed at Karataş, especially when compared with other Eastern Mediterranean populations dated between the Neolithic and the Middle Bronze Age. Adams (2020) stated that the Anatolian populations included in the study for comparative purposes (Laodikeia and Çine Tepecik) were more similar to each other than their Eastern Mediterranean contemporaries, but that a diversity existed regarding female migration. On the basis of the significant metric and non-metric variation in Karataş females, a similar (to that suggested by Pilloud et al. (2017) for Çatalhöyük, Aşıklı and Musular) patrilocal post-marital settlement practice likely existed at Karataş. In another study investigating the biological relations of western Anatolian populations (Akbaba 2020), the dental morphological traits of 303 individuals belonging to western (Batı Limanı, İzmir Demir Çelik Limanı and Klazomenai) and central (Topaklı) Anatolian populations were analysed. Akbaba (2020) found that populations dating to the earlier periods and located in the same region had greater biological affinity. He also compared his data with those of previously studied Anatolian populations (Resuloğlu, İkiztepe, İznik and Cevizcioğlu populations with those of Boncuklu, Aşıklı Höyük, Musular and Çatalhöyük), and with other parts of the world. He concluded that populations of the earlier periods clustered closer together, while those of the later periods did not cluster together as closely.

By analysing the frequency distribution of 17 non-metric traits of 37 Sagalassos individuals and 27 Eurasian and African populations dated between the 11th and 13th centuries AD, Ricaut and Waelkens (2008) revealed different layers of biological history at Sagalassos. A biological affinity between the Sagalassos and Eastern Mediterranean populations was established, as well as similarities with Sub-Saharan and Northern and Central European populations (Ricaut, Waelkens 2008). The results of mtDNA variation analysis carried out on the same skeletal population indicated a biological integration with West Eurasian populations (Ottoni et al. 2011). The combination of the biological anthropological data with historical and archaeological data has allowed the researchers to state that genetic relationships existed between Balkan and Greek populations at Sagalassos, as well as between populations of ancient Persia, the Italian Peninsula and the Levant, while there were no genetic contributions from Central Asian populations (Ottoni et al. 2011). The results of this study would appear to be compatible with those of other genetic and anthropological investigations (Ricaut, Waelkens 2008) performed on Anatolian skeletal populations.

In a study analysing non-metric data of 11 Anatolian populations unearthed from different areas of the peninsula dated to the Middle Ages, it was found that diversity was higher in this period when compared to the previous periods (Eroğlu 2019). When all the groups were combined, the Anatolian populations clustered together with European, West Asian and North African populations. A similar result was reached by comparing supraorbital torus and hypoglossal canal traits of the same populations with those of 72 populations worldwide (Eroğlu 2016).

Discussion

Anatolia, which was inhabited by modern humans from the Lower Palaeolithic to the Modern Age (Kuhn 2002) and served as a land bridge for migrations between the Caucasus, Asia, Europe and Africa, thus acted as a bridge for gene flow as well. Migration has had a very significant cultural and evolutionary impact. The genetic makeup of the populations of Anatolia, which has historically hosted many civilisations, has changed over time. Anthropological studies dated to both before (Krogman 1937; 1949; Şenyürek 1941; 1951; 1952; Mellink, Angel 1968; Cappieri 1969; 1970; 1972; Clement et al. 1974; Özbek 1990) and after 1980 (Wittwer-Backofen 1985; 1987; Özer, Güleç 2000; Açıkkol et al. 2005; Eroğlu, Erdal 2006; Pilloud 2009; Seguchi et al. 2010; Adams 2020) have demonstrated that the inhabitants of Anatolia displayed a homogenous structure in the Neolithic period. This pattern was discovered following the analysis of 24 craniofacial traits of 52 populations from Anatolia and its neighbouring regions (Brace et al. 2006), and has been supported by archaeological (Bilgi 2001; Sagona, Zimansky 2015; Frangipane 2017) and aDNA (Mathieson 2015; Kılınç et al. 2016; Omrak et al. 2016; Feldman et al. 2019; Skourtanioti et al. 2020) studies as well. Undoubtedly, one the most important developments in the Neolithic period was the adoption and spread of a lifestyle based on agriculture. When investigating whether the spread of this agricultural way of life from Anatolia to Europe was 'culturally' or 'demic'-based, 'demic diffusion' has become the dominant view (Özdoğan 2011; Lazaridis et al. 2014). The spread of Neolithisation within and outside the Anatolian peninsula occurred in two stages (Özdoğan 2011). According to Özdoğan (2011), the spread of the Neolithic system was the result of movement through migration and colonisation via both land and sea. The first stage, characterised by slow and unorganised movement, continued until the second half of the seventh millennium, while the immediately succeeding second stage had a faster spread than the first one; proof of this can be found in western Anatolian populations (e.g. Çukuriçi Höyük, Ulucak and Yeşilova).

Anatolia had a key position in the spread of agriculture, which started in Mesopotamia ca 10,000–9000 BC, to Europe (Baird et al. 2018). The primary Neolithisation site, which was long thought to be restricted to the Fertile Crescent and the surrounding areas, has now been expanded to cover the southeastern parts of the central Anatolia basin (Özdoğan 2011). Evidence for the first emergence of agriculture in central Anatolia in the second half of the ninth millennium BC comes from two central Anatolian settlements, the hunter-gatherer site of Pınarbaşı and the agriculturalist settlement of Boncuklu, both characterised by the use of similar chipped stone tools. These settlements, which have evidence for human occupation and activity from the Epipalaeolithic to the Early Holocene (9800–7800 BC), were found to be contemporary with the Levantine Natufians, who practised agriculture (Baird et al. 2018). Ancient DNA analyses of these populations (Feldman et al. 2019) showed that the Neolithic Anatolian populations inherited a large part of their lineage from those of the Epipalaeolithic. Therefore, it is possible to suggest that agriculture was adopted locally by the hunter-gatherers of central Anatolia. Similarly, genetic studies have shown that the populations of Iran and the Levant originated from local hunter-gathering groups of the Early Neolithic period (Broushaki et al. 2016; Feldman et al. 2019; Skourtanioti et al. 2020). In addition to the local genetic contribution of the early Anatolian populations, a Neolithic Iranian/Caucasian influence has been observed in the genes of Anatolian aceramic agriculturists, while the Neolithic Levantine populations genetically affected the Anatolian ceramic agriculturists (Feldmand et al. 2019). After the Neolithic period, and especially in the Chalcolithic, the existence of a morphological (Özer, Güleç 2000; Açıkkol et al. 2005; Eroğlu, Erdal 2006; Ricaut, Waelkens 2008; Seguchi et al. 2010; Eroğlu 2011; 2016; 2019; Adams 2020; Akbaba 2020) and cultural (Sahoğlu 2005; Sagona, Zimansky, 2015; Frangipane 2017; Rothman 2017) heterogeneity in Anatolia, as a result of migrations, has been accepted. Genetic data has also demonstrated that there was population mobility from the east and south to the west from the Late Chalcolithic period and throughout the Bronze Age (Lazaridis et al. 2016; 2017; Harney et al. 2018; Skourtanioti et al. 2020). Skourtanioti et al. (2020) have identified a gradual genetic change and drift from the southern Caucasus towards western Anatolia during the Late Neolithic / Early Chalcolithic periods. This gradual genetic change extended into the Chalcolithic / Bronze Age from Central Asia to the Zagros region, which included a small part of eastern Anatolia. It has been observed that genetic changes that occurred during the Bronze Age in West Asia did not have a different genetic contribution due to outside genetic incursions, but rather were the result of internal dynamics originating from the genetic combination that took place in the Late Neolithic and Early Chalcolithic periods. According to Lazaridis et al. (2017), the Bronze Age populations of the Aegean region were similar to each other and to southwestern Bronze Age Anatolians, also showing eastern influences from the Caucasus. A similar situation was put forward by Mellink and Angel (1968). Thus, the study of Skourtanioti et al. (2020) suggesting that Chalcolithic and Bronze Age populations from western and central Anatolia, the southern Levant, Iran (Zagros) and Caucasus showed less genetic differentiation almost two thousand years after the beginning of agriculture – and who characterised the gene flow process across a large area in later periods – provides archaeogenomic evidence for the clustering of Anatolian Bronze Age populations close to each other and to populations to the south of Anatolia (Eroğlu, Erdal 2006). Moreover, it has been suggested that West Eurasian populations, dated from the Late Chalcolithic period to the Late Bronze Age, had a common genetic profile that may have originated from the Late Neolithic–Early Chalcolithic lineages (Skourtanioti et al. 2020).

Archaeological and historical data (Köprülü 1980; Uzunçarşılı 1981; Boardman et al. 1992) indicate that Anatolia has been subject to migrations, invasions and conflict since the Bronze Age. Traces of possible genetic contribution from the Middle East, Central Asia and Siberia to contemporary Anatolian groups have also been suggested (Omrak et al. 2016). Large-scale population movements, from the south Caucasus and Near Eastern and Turkic migrations, have contributed to the genetic makeup of contemporary Anatolian populations (Omrak et al. 2016). These traces have been observed since the Middle Ages (Ricaut, Wilkens 2008; Eroğlu 2019), when gene flow appears to be most rigorous compared to previous periods (Boardman et al. 1992; Eroğlu 2019). Historical data shows that from the 11th century AD, there was a regular genetic flow towards Anatolia from Central Asia (Köprülü 1981; Uzunçarşılı 1982). Investigations on cranial and dental morphological traits also indicate that these traits, related to Asian populations, increased after this point (Erdal 1992; Erdal, Eroğlu, 2000; Eroğlu, Erdal 2008; Erdal 2011; Eroğlu 2012). The frequency of palatine torus, observed in ancient Anatolian skeletal populations from the Early Bronze Age to the first quarter of the 20th century, reflects a Mongoloid gene flow (Eroğlu, Erdal, 2008). The frequency of palatine torus was particularly high during the Ottoman period. This data is supported by a recent DNA study conducted on modern living people in Anatolia from four Turkish cities (Ankara, Antalya, Izmir and Van) (Di Benedetto et al. 2001). This study demonstrated the presence of a continuous gene flow from Central Asia to Anatolia since the 11th century AD. Anatolian populations dating to the 20th century (Erzurum and Hakmehmet) showed a lower frequency of palatine torus compared to the earlier populations. This situation indicates that the gene flow into Anatolia continued in the 20th century as well, although the direction changed, becoming an eastwards flow from Europe to Anatolia.

The frequency of metopism (Eroğlu 2008) and hypoglossal canal bridging (Eroğlu 2010; 2016) observed in the same populations where palatine torus was detected is similar to that found in European populations (Eroğlu 2008). Ricaut and Waelkens (2008) found a low biological affinity between Sagalassos and Central Asian populations. Seguchi et al. (2010) studied Turkish population substructure and history by examining craniofacial diversity across several temporal periods framed within a population genetic model. Their results indicated minimal Turkic Central Asian genetic influence on the Anatolian population structure.

Similarly, most genetic studies on people living in Türkiye suggest that current Anatolian populations are closer to European populations than Central Asian populations (Calafell et al. 1996; Comas et al. 1996; Rolf et al. 1999; Richards et al. 2000; Cinnioglu et al. 2004; Berkman et al. 2008; Gonzalez-Perez et al. 2010). Comas et al. (1996) analysed the hypervariable segment 1 control sequence in the mtDNA of 45 unrelated individuals from Anatolia. This study suggests that a demographic expansion occurred in the Middle East, spreading from Türkiye into Europe. After an initial expansion, high levels of migration (gene flow) during the Neolithic period must have contributed to a homogenisation of the genetic landscape of Europe. Calafell et al. (1996) found similar results in their study of 30 Bulgarians and 29 Turks. Cinnioglu et al. (2004), in an analysis of biallelic polymorphisms for Turkish Y chromosomes, noticed that major haplogroups (combination of linked alleles at linked loci sharing a common ancestral mutation) were mostly shared with European and Middle Eastern populations (94.1%), followed by Central Asian (3.4%), Indian (1.5%) and African (1%) haplogroups. They also identified detectable, yet weak (<9%), signals of paternal gene flow from Central Asia. King et al. (2008) found similar results. Another study based on the analyses of six STR loci in 88 Y-chromosomes from Türkiye (Rolf et al. 1999) suggested only a 10% contribution from Central Asian populations. By using Alu insertion polymorphisms, Berkman et al. (2008) recognised a similar Central Asian genetic contribution to Anatolia, estimating ca 13% admixture. Gonzalez-Perez et al. (2010) analysed the biological/genetic relationships between human groups in 15 Mediterranean populations, including Türkiye. They examined the variation of 18 Alu polymorphisms and three linked STRs. They found a close relationship between Türkiye and Greece. Also, mtDNA studies on European and Middle Eastern populations had similar results (Richards et al. 2000).

Conclusion

In the field of biological distance studies in Anatolia, as well as globally, there has been a more recent transition from a typological approach to those which have, instead, adopted a population-oriented approach. However, these current studies mark the beginning of a new phase for Anatolia. The reconstruction of the history and genetic makeup of the populations that have lived in Anatolia, which has hosted hundreds of civilisations, is an arduous and long process. Together with archaeological and anthropological studies, the role of aDNA studies, especially in the last years, has been very important in illuminating this field of research. In particular, the fact that recent osteological biological distance studies (Eroğlu, Erdal, 2006; Ricaut, Wealkens, 2008; Pilloud, Larsen, 2011) have yielded results in parallel with those of aDNA studies (Ottoni et al. 2011; Skourtanioti et al. 2020; Yaka et al. 2021) has demonstrated the continued importance of utilising skeletal morphological data when working towards the reconstruction and comprehension of the population history of Anatolia.

Unfortunately, many archaeological populations in Türkiye have not been adequately studied in terms of biological distance. Özdoğan (2011) highlighted that out of a total of 26 Neolithic sites excavated in the previous 20 years in Türkiye, 13 were still being excavated as of 2011. Therefore, many skeletal populations are still waiting to be studied and analysed. Another important issue in biological distance studies is methodology. It should be emphasised that a standardised methodology should be established and used for comparing the data.

In order to better understand diachronic population mobility and its dynamics in Anatolia and in the surrounding regions, each chronological period should be individually studied. It has also been observed that a multidisciplinary approach, including the investigation of cranial and dental metric traits, as well as aDNA and isotope analyses, is much more effective.

Bibliography

Açıkkol, A., Yılmaz, H., Özer, İ., Sağır, M., Güleç, E. 2005: 'Eski Anadolu toplumlarının kladistik analizi' *Arkeometri Sonuçları Toplantısı* 20: 113–24

Adams, D.M. 2020: *Community Development and Culture Negotiation during the Early Bronze Age in Western Anatolia*. PhD thesis, The University of Nevada, Reno

Baird, D., Fairbairn, A., Jenkins, E., Martin, L., Middleton, C., Pearson, J., Asouti, E., Edwards,Y., Kabukcu, C., Mustafaoğlu, G., Russell, N., Bar-Yosef, O., Jacobsen, G., Wu, X. Baker, B., Elliott, S. 2018: 'Agricultural origins on the Anatolian plateau' *Proceedings of the National Academy of Sciences* 115.14: E3077–E3086

Berkman, C.C., Dinc, H., Sekeryapan, C., Togan, I. 2008: 'Alu insertion polymorphisms and an assessment of the genetic contribution of Central Asia to Anatolia with respect to the Balkans' *American Journal of Physical Anthropology* 136: 11–18

Bilgi, Ö. 1990: 'Metal objects from İkiztepe-Turkey' *Beiträge zur Allgemeinen und Vergleichenden Archäologie* 9.10: 119–219

— 2001: Çağda Orta Karadeniz Bölgesi Madenciliği; Hint-Avrupalıların Anavatanı Sorununa Yeni Bir Yaklaşım TASK Vakfı Yayınları. 4, Monografi Serisi 1, İstanbul

— 2003: 'İkiztepe mezarlık kazıları ve ölü gömme gelenekleri 2000 ve 2001 dönemleri' *Belleten* 67.249: 383–402

Boardman, J., Edwards, I.E.S., Sollberger, E., Hammond, N.G.L. (eds) 1992: *The Cambridge Ancient History*. Cambridge, Cambridge University Press

Brace, C.L. 2010: '"Physical" anthropology at the turn of the last century' in M.A. Little, K.A.R. Kennedy (eds), *Histories of American Physical Anthropology in the Twentieth Century*. New York, Rowman and Littlefield Publishers: 25–54

Brace, C.L., Seguchi, N., Quintyn, C.B., Fox, S.C., Nelson, A.R., Manolis, S.K., Qifeng, P. 2006: 'The questionable contribution of the Neolithic and the Bronze Age to European craniofacial form' *Proceedings of the National Academy of Sciences* 103.1: 242–47

Buikstra, J.E. 1977: 'Biocultural dimensions of archaeological study: A regional perspective' in R.L. Blakely (ed.), *Biocultural Adaptation in Prehistoric America*. Athens, GA, University of Georgia Press: 67–84

— 1979: 'Contributions of physical anthropologists to the concept of Hopewell: a historical perspective' in D.S. Brose, N. Greber (eds), *Hopewell Archaeology: The Chillicothe Conference*. Kent, OH Kent State University Press: 220–33

Buikstra, J.E., Frankenberg, S.R., Konigsberg, L.W. 1990: 'Skeletal biological distance studies in American physical anthropology: recent trends' *American Journal of Physical Anthropology* 82.1: 1–7

Calafell, F., Underhill, P., Tolun, A., Anglicheva, D., Kalaydjieva, L. 1996: 'From Asia to Europe: mitochondrial DNA sequence variability in Bulgarian and Turks' *Annals of Human Genetics* 60: 35–49

Cappieri, M., 1969: 'The Mediterranean race in Asia before the Iron Age' in H. Field (ed.), *Occasional Paper 8*. Miami, Field Research Projects: 1–28

— 1970: 'The Anatolians of the late Neolithic and Chalcolithic Age' *Belleten* 34: 509–58

— 1972: 'Genotypical and phenotypical homogeneity of the ancient population of Anatolia' in H. Field (ed.), *Occasional Paper 63*, Miami, Field Research Projects: 25–40

Carson, E.A. 2006a: 'Maximum-likelihood variance components analysis of heritabilities of cranial nonmetric traits' *Human Biology* 78.4: 383–402

— 2006b: 'Maximum likelihood estimation of human craniometric heritabilities' *American Journal of Physical Anthropology* 131.2: 169–80

Celement, F.M., Dabagihan, E., Ferembach, D., Jouanneau, J. 1974: 'Les alpins de Çatal Höyük (Néolithique, Turquie). Application d'une analyse en composante principale normée á la définition de leur origine' *Biométrie Humaine* (Paris) 9: 53–69

Cheverud, J.M., Buikstra, J.E. 1981: 'Quantitative genetics of skeletal nonmetric traits in the rhesus macaques on Cayo Santiago. I. Single trait heritabilities' *American Journal of Physical Anthropology* 54.1: 43–49

Cheverud, J.M., Buikstra, J.E. 1982: 'Quantitative genetics of skeletal nonmetric traits in the rhesus macaques of Cayo Santiago. III. Relative heritability of skeletal nonmetric and metric traits' *American Journal of Physical Anthropology* 59.2: 151–55

Çiner, R. 1964. 'Bodrum-Müskebi kazısı iskelet kalıntılarının tetkiki' *Antropoloji* 1.2: 56–79

Cinnioğlu, C., King, R., Kivisild, T., Kalfoğlu E., Atasoy, S., Cavalleri, G.L., Lillie, A.S., Roseman, C.C., Lin, A.A., Prince, K., Oefner, P.J., Shen, P., Semino, O., Cavalli-Sforza, L.L., Underhill, P.A. 2004: 'Excavating Y-chromosome haplotype strata in Anatolia' *Human Genetics* 14: 127–48

Comas, D., Calafell, F., Mateu, E., Pérez-Lezaun, A., Bertranpetit, J. 1996: 'Geographic variation in human mitochondrial DNA control region sequence: the population history of Turkey and its relationship to the European populations' *Molecular Biology and Evolution* 13: 1067–77

Di Benedetto, G.D., Güven, A., Stenico, M., Bertonelle, G. 2001: 'DNA diversity and population admixture in Anatolia' *American Journal of Physical Anthropology* 115: 144–56

Erdal, Y.S. 1992: 'İznik Geç Bizans dönemi insanlarının kafatası yapıları' *Türk Arkeoloji Dergisi* 30: 51–69

— 2011: 'Tasmasor Geç Demir Çağı iskeletlerinin antropolojik analizi' in S.Y. Şenyurt (ed.) *Tasmasor*. Ankara, Bilgin Sanat Yayınları: 329–458

Erdal, Y.S., Eroğlu, S. 2000: 'Bazı Anadolu topluluklarında diş varyasyonlarının dağılımı' 2. *Ulusal Biyolojik Antropoloji Sempozyumu Ankara: October 25–26*

Eroğlu, S. 2008. 'The frequency of metopism in Anatolian populations dated from the Neolithic to the first quarter of the 20th century' *Clinical Anatomy* 21: 471–78

— 2009: 'Ölçülemeyen diş özelliklerinin biyolojik uzaklık çalışmalarındaki önemi' *Mustafa Kemal Üniversitesi Sosyal Bilimler Enstitüsü Dergisi* 6.11: 543–67

— 2011: 'Comparison of cranial metric and non-metric traits in the determining of biological distance: an example of Anatolia' TÜBA-AR-Turkish Academy of Sciences Journal of Archaeology 14: 231–44

— 2016: 'Supraorbital foramen and hypoglossal canal bridging in ancient/modern Anatolian populations: implications for worldwide population distribution' *Mediterranean Archaeology and Archaeometry* 16.1: 169–83

Eroğlu, S., Erdal, Y.S. 2006: 'İkiztepe insanlarının Anadolu ve Anadolu'ya yakın topluluklarla biyolojik ilişkisi' *Arkeometri Sonuçları Toplantısı* 21: 41–50

— 2008: 'Why did the frequency of palatine torus increase in the ancient Anatolian populations?' *HOMO – Journal of Comparative Human Biology* 59.5: 365–82

— 2009: 'Diş ve kafatası morfolojisine dayanarak üç eski Anadolu topluluğunda biyolojik uzaklıkların belirlenmesi' *Hacettepe Diş Hekimliği Fakültesi Dergisi* 33.3: 78–90

Feldman, M., Fernández-Domínguez, E., Reynolds, L., Baird, D., Pearson, J., Hershkovitz, I., May, H., Goring-Morris, N., Benz, M., Gresky, J., Bianco, R.A., Fairbairn, A., Mustafaoğlu, G., Stockhammer, P.W., Posth, C., Haak, W., Jeong, C., Krause, J. 2019: 'Late pleistocene human genome suggests a local origin for the first farmers of central Anatolia' *Nature Communications* 10.1: 1218

Ferembach, D. 1972: 'Les hommes du gisement Neolithique de Çatal Hüyük' *VII Turk Tarih Kongresi* (1970), Ankara, Türk Tarih Kurumu Kongresi Yayınları: 15–21

Frangipane, M. 2017: 'The role of metal procurement in the wide interregional connections of Arslantepe during the late 4th–early 3rd millennia BC' in Ç. Maner, M.T. Horowitz, A.S. Gilbert (eds), *Overturning Certainties in Near Eastern Archaeology*, Boston, Brill: 186–210

Herrera, B., Hanihara, T., Godde, K. 2014: 'Comparability of multiple data types from the Bering Strait region: cranial and dental metrics and nonmetrics, mtDNA, and Y-chromosome DNA' *American Journal of Physical Anthropology* 154.3: 334–48

Hubbard, A.R., Guatelli-Steinberg, D., Irish, J.D. 2015: 'Do nuclear DNA and dental nonmetric data produce similar reconstructions of regional population history? An example from modern coastal Kenya' *American Journal of Physical Anthropology* 157.2: 295–304

Kansu, Ş., Atasayan, A.M. 1939: 'Afyonkarahisar civarında kusura afriyatında Meydana Çıkarılan Bakır Çağı ve Eti Devrine ait iskeletler üzerinde tetkikler' *Türk Antropoloji Mecmuası*: 19–22

Kılınç, G.M., Omrak, A., Özer, F., Günther, T., Büyükkarakaya, A.M., Bıçakçı, E., Baird, D., Dönertaş, H.M., Ghalichi, A., Yaka, R., Koptekin, D., Açan, S.C., Parvizi, P., Krzewińska, M., Daskalaki, E.A., Yüncü, E., Dağtaş, N.D.,

Fairbairn, F., Götherström, A. 2016: 'The demographic development of the first farmers in Anatolia' *Current Biology* 26.19: 2659–66

Klung, S.T. Witwer-Backofen, U. 1983: 'Diskreta im populations vergleich' *Homo* 34.3–4: 153–68

Knudson, K.J., Stojanowski, C.M. 2008: 'New directions in bioarchaeology: Recent contributions to the study of human social identities' *Journal of Archaeological Research* 16.4: 397–432

Köprülü, F. 1981: *Osmanlı İmparatorluğunun Kuruluşu*. İstanbul, Ötüken Yayınevi

Krogman, W.M. 1937: 'Cranial types from Alishar Hüyük and their relations to other racial types, ancient and modern of Europe and Western Asia' in H.H. Von der Osten (ed.) *The Alishar Hüyük, Seasons of 1930–32*, Part III, Chicago, The Oriental Institute 30: 213–93

— 1949: 'Ancient cranial types at Chatal Hüyük and Tell al-Judaidah' Syria, from the late fifth millennium BC to the mid-seventh century AD' *Belleten* 51: 407–533

Kuhn, S.L. 2002: 'Paleolithic archeology in Turkey' *Evolutionary Anthropology* 11: 198–210

Larsen, C.S. 2002: 'Bioarchaeology: the lives and lifestyles of past people' *Journal of Archaeological Research* 10.2: 119–66

Larsen, C.S., Knüsel, C.J., Haddow, S.D., Pilloud, M.A., Milella, M., Sadvari, J.W., Pearson, J., Ruff, C.B., Garofalo, E.M., Bocaege, E., Betz, B.J., Dori, I., Glencross, B. 2019: 'Bioarchaeology of Neolithic Çatalhöyük reveals fundamental transitions in health, mobility, and lifestyle in early farmers' *Proceedings of the National Academy of Sciences* 116.26: 12615–23

Lazaridis, I., Nadel, D., Rollefson, G., Merrett, D.C., Rohland, N., Mallick, S., Fernandes, D., Novak, M., Gamarra, B., Sirak, K., Connell, S., Stewardson, K., Harney, E., Fu, Q., Gonzalez-Fortes, G., Jones, E.R., Alpaslan Roodenberg, S., Lengyel, G., Bocquentin, F., Gasparian, B., Monge, J.M., Gregg, M., Eshed, V., Mizrahi, A.S., Meiklejohn, C., Gerritsen, F., Bejenaru, L., Blüher, M., Campbell, A., Cavalleri, G., Comas, D., Froguel, F., Gilbert, E., Kerr, S.M., Kovacs, P., Krause, J., McGettigan, D., Merrigan, M., Merriwether, D.A., O'Reilly, S., Richards, M.B., Semino, O., Shamoon-Pour, M., Stefanescu, G., Stumvoll, M., Tönjes, A., Torroni, A., Wilson, J.F., Yengo, L., Hovhannisyan, N.A., Patterson, N., Pinhasi, R., Reich, D. 2016: 'Genomic insights into the origin of farming in the ancient Near East' *Nature* 536.7617: 419–24

Lazaridis, I., Mittnik, A., Patterson, N., Mallick, S., Rohland, N., Pfrengle, S., Furtwängler, A., Peltzer, A., Posth, C., Vasilakis, A., McGeorge, P.J.P., Konsolaki-Yannopoulou, E., Korres, G., Martlew, H., Michalodimitrakis, M., Özsait, M., Özsait, N., Papathanasiou, A., Richards, M., Alpaslan Roodenberg, S., Tzedakis, Y., Arnott, R., Fernandes, D.M., Hughey, J.R., Lotakis, D.M., Navas, P.A., Maniatis, Y., Stamatoyannopoulos, J.A., Stewardson, K., Stockhammer, P., Pinhasi, R., Reich, D., Krause, J., Stamatoyannopoulos, G. 2017: 'Genetic origins of the Minoans and Mycenaeans' *Nature* 548.7666: 214–18

Little, M.A., Sussman, R.W. 2010. 'History of biological anthropology' in C.S. Larsen (ed.), *A Companion to Biological Anthropology*. Singapore, John Wiley and Sons: 13–38

Mathieson, I., Lazaridis, I., Rohland, N., Mallick, S., Patterson, N., Roodenberg, S.A., Harney, E., Stewardson, K., Fernandes, D., Novak, M., Sirak, K., Gamba, C., Jones, E.R., Llamas, B., Dryomov, S., Pickrell, J., Arsuaga, J.L., Bermúdez de Castro, J.M., Carbonell, E., Gerritsen, F., Khokhlov, A., Kuznetsov, P., Lozano, M., Meller, H., Mochalov, O., Moiseyev, V., Rojo Guerra, M.A., Roodenberg, J., Vergès, J.M., Krause, J., Cooper, A., Alt, K.W., Brown, D., Anthony, A., Lalueza-Fox, C., Haak, W., Pinhasi, R., Reich, R. 2015: 'Genome-wide patterns of selection in 230 ancient Eurasians' *Nature* 528: 499–503

Mellink, M.J., Angel, J.L. 1968: 'Excavations at Karataş-Semayük in Lycia, 1967' *American Journal of Archaeology*: 243–63

Molnar, S. 2015: *Human Variation: Races, Types, and Ethnic Groups*. New York, Routledge

Nikita, E. 2017: *Osteoarchaeology: A Guide to the Macroscopic Study of Human Skeletal Remains*. London, Academic Press

— 2020: 'Biodistance studies in the Eastern Mediterranean and Middle East: an overview and future prospects' *Journal of Archaeological Science*: Reports 34: 102647

O'Rourke, D., Enk, J. 2012: 'Genetics, geography, and human variation' in S. Stinson, B. Bogin, D. O'Rourke (eds) *Human Biology: An Evolutionary and Biocultural Perspective*. Hoboken, John Wiley and Sons, Inc.: 99–141

Omrak, A., Günther, T., Valdiosera, C., Svensson, E.M., Malmström, H., Kiesewetter, H., Götherström, A. 2016: 'Genomic evidence establishes Anatolia as the source of the European Neolithic gene pool' *Current Biology* 26.2: 270–75

Ottoni, C., Ricaut, F.X., Vanderheyden, N., Brucato, N., Waelkens, M., Decorte, R. 2011: 'Mitochondrial analysis of a Byzantine population reveals the differential impact of multiple historical events in south Anatolia' *European Journal of Human Genetics* 19.5: 571–76

Özbek, M. 1988: 'Çayönü insanları ve sağlık sorunları' *IV. Arkeometri Sonuçları Toplantısı* 4: 121–52

— 1990: 'Anadolu'da insan ırklarının tarihsel gelişimi' *Türk Tarih Kongresi* 11: 45–68

— 1992: 'Aşıklı Höyük Neolitik Çağ insanları' *Arkeometri Sonuçları Toplantısı* 8: 201–12

— 1993: 'Aşıklı Höyük insanları (1990 ve 1992 buluntuları)' *Arkeometri Sonuçları Toplantısı* 9: 23–32

— 2001: 'Cumhuriyetle Gelişen antropoloji' *Atatürk'ün Ölümünün 62. Yılında Cumhuriyet Türkiye'sinde Bilimsel Gelişmeler Sempozyumu, 8–10 Kasım 2000*. Ankara, Hacettepe Üniversitesi Atatürk İlkeleri ve İnkılap Tarihi Enstitüsü Yayını Ankara: 119–26

Philbin, C.S., Pilloud, M.A. 2018: 'Dental diversity and population movement in Neolithic central Anatolia' *Program of the 87th Annual Meeting of the American Association of Physical Anthropologists*, April 11–14, 165: 206–07

Pilloud, M.A. 2009: *Community Structure at Neolithic Çatalhöyük: Biological Distance Analysis of Household, Neighborhood, and Settlement*. PhD thesis, The Ohio State University, Columbus

Pilloud, M.A., Hefner, J.T. (eds) 2016: *Biological Distance Analysis: Forensic and Bioarchaeological Perspectives*. San Diego, Academic Press

Pilloud, M.A., Somel, M., Haddow, S.D., Knusel, C.J., Larsen, C.S., Özbaşaran, M., Erdal, Ö.D., Baird, D., Pearson, J., Götherström, A., Storå, J., Jakobsson, M., Kılınç, G.M., Özer, F., Koptekin, D., Dagtas, N.D. 2017: 'Mobility in Neolithic central Anatolia: a comparison of dental morphometrics and aDNA' *American Journal of Physical Anthropology* 162: 316

Relethford, J.H. 1996: 'Genetic drift can obscure population history: problem and solution' *Human Biology*: 29–44

Ricaut, F.X., Waelkens, M. 2008: 'Cranial discrete traits in a Byzantine population and eastern Mediterranean population movements' *Human Biology* 80.5: 535–64

Ricaut, F.X., Auriol, V., von Cramon-Taubadel, N., Keyser, C., Murail, P., Ludes, B. Crubézy, E. 2010: 'Comparison between morphological and genetic data to estimate biological relationship: the case of the Egyin Gol necropolis (Mongolia)' *American Journal of Physical Anthropology* 143.3: 355–64

Rolf, B., Röhl, A., Forster, P., Brinkmann, B. 1999: 'On the genetic origins of the Turks: study of six Y-chromosomal short tandem repeats' in S. Papiha, R. Deka, R. Chakraborty (eds), *Genomic Diversity: Applications in Human Population Genetics*. New York, Kluwer Academic / Plenum: 75–82

Rothman, M.S. 2017: 'Explaining the Kura-Araxes' in K.O. Weber, E. Hite, L. Khatchadourian, A.T. Smith (eds), *Fitful Histories and Unruly Publics: Rethinking Temporality and Community in Eurasian Archaeology*. Leiden, Brill: 217–57

Sagona, A.G., Zimansky, P. 2015: *Arkeolojik Veriler Işığında Türkiye'nin En Eski Kültürleri: MÖ 1.000.000–550*. İstanbul, Arkeoloji ve Sanat Yayınları

Şahoğlu, V. 2005: 'The Anatolian trade network and the Izmir region during the Early Bronze Age' *Oxford Journal of Archaeology* 24.4: 339–61

Scott, G.R., Turner, C.G. 1997: *Anthropology of Modern Human Teeth*. Cambridge, Cambridge University Press

Self, S.G., Leamy, L. 1978: 'Heritability of quasi-continuous skeletal traits in a randombred population of house mice' *Genetics* 88.1: 109–20

Şenyürek, M.S. 1941: 'Anadolu Bakır Çağı ve Eti Sekenesinin kraniyolojik tetkiki' *Belleten* 19: 219–53

— 1951: 'Fluctuation of the cranial index in Anatolia, from the fourth millennium BC to 1200 BC' *Belleten* 15.38: 593–615

— 1952: 'A study of the human skeletons from Kültepe excavated under the auspices of the Turkish Historical Society: the skeletons from the excavation season of 1948' *Belleten* 63: 323–43

Sjøvold, T. 1984: 'A report on the heritability of some cranial measurements and non-metric traits' in G.N. Van Vark, W.W. Howells (eds), *Multivariate Statistical Methods in Physical Anthropology*. Dordrecht, Springer: 223–46

Skourtanioti, E., Erdal, Y.S., Frangipane, M., Restelli, F.B., Yener, K.A., Pinnock, F., Matthiae, P., Özbal, R., Schoop, U.-D., Guliyev, F., Akhundov, T., Lyonnet, B., Hammer, E.L., Nugent, S.E., Burri, M., Neumann, G.U., Penske, S., Ingman, T., Akar, M., Shafiq, R., Palumbi, G., Eisenmann, S., D'Andrea, M., Rohrlach, A.B.,Warinner, C., Jeong, C., Stockhammer, P.W., Haak, W., Krause, J. 2020: 'Genomic history of Neolithic to Bronze Age Anatolia, Northern Levant, and Southern Caucasus' *Cell* 181.5: 1158–75

Stini, W.A. 2010: 'Sherwood Washburn and the "new physical anthropology"' in M.A. Little, K.A.R. Kennedy (eds), *Histories of American Physical Anthropology in the Twentieth Century*. New York, Rowman and Littlefield: 173–86

Stojanowski, C.M., Schillaci, M.A. 2006: 'Phenotypic approaches for understanding patterns of intracemetery biological variation' *American Journal of Physical Anthropology* 131.43: 49–88

Susanne, C. 1975: 'Genetic and environmental influences on morphological characteristics' *Annals of Human Biology* 2.3: 279–87

— 1976: 'Heredity of anthropometric measurements: analysis with the method of Fisher (1918)' *Yugoslav Anthropological Society* 13: 11–20

Thissen, L. 1993: 'New insights in Balkan-Anatolian connections in the Late Chalcolithic: old evidence from the Turkish Black Sea littoral' *Anatolian Studies* 43: 207–37

Uzunçarşılı, İ.H. 1982: *Osmanlı Tarihi*. Ankara, Türk Tarihi Kurumu Yayınları

Welton, M.L. 2010: *Mobility and social organization on the ancient Anatolian Black Sea coast: an archaeological, spatial and isotopic investigation of the cemetery at İkiztepe, Turkey*. PhD thesis, University of Toronto, Toronto

Williams, S.R., Chagnon, N.A., Spielman, R.S. 2002: 'Nuclear and mitochondrial genetic variation in the Yanomamö: a test case for ancient DNA studies of prehistoric populations' *American Journal of Physical Anthropology* 117.3: 246–59

Wittwer-Backofen, U. 1985: 'Anthropologische untersuchungen der Nekropole Ikiztepe/Samsun' *Araştırma Sonuçları Toplantısı* 3: 421–28

Yaka, R., Mapelli, I., Kaptan, D., Doğu, A., Chyleński, M., Erdal, Ö.D., Koptekin, D., Vural, B., Bayliss, A., Mazzucato, C., Fer, E., Çokoğlu, S.S., Lagerholm, K., Krewinska, M., Karamurat, C., Gemici, C.G., Sevkar, A., Dağtaş, N.D., Kılınç, G.M., Adams, D., Murters, A.R., Sağlıcan, E., Milella, M., Schotsmans, E.M.J., Yurtman, E., Çetin, M., Yorulmaz, S., Altınışık, E., Ghalichi, A., Bilgin, C., Günther, T., Stora, J., Jakobsson, M., Kleijn, M., Mustafaoğlu, G., Fairbairn, A., Pearson, J., Togan, İ., Kayacan, N., Marciniak, A., Larsen, C.S., Hodder, I., Atakuman, Ç., Pilloud, M., Sürer, E., Erritsen, F.G., Özbal, R., Baird, D., Erdal, Y.S., Duru, G., Özbaşaran, M., Haddow, D., Knüsell, C.J., Götherström, A., Özer, F., Somel, M. 2021: 'Variable kinship patterns in Neolithic Anatolia revealed by ancient genomes' *Current Biology* 31: 2455–68

6. Intentional headshaping in Anatolia. A diachronic analysis of its typological and semantic distribution

Valentina D'Amico

Introduction

Since prehistory the skull has been conceived as the most important part of the human body. Ethnographic research, archaeological findings and anthropological investigations have confirmed its importance among different cultures worldwide, and in different chronological periods (Bonogofsky 2011). In the history of anthropological studies, for a long time, the shape and size of the skull and, more generally, the morphology of the human body were considered static and unchangeable. That is why the skull was accepted as the body component best reflecting the genetic makeup of an individual. However, new studies have superseded the old theories in favour of human body plasticity and flexibility derived from both genetic and environmental factors (Hauser, De Stefano 1989). Therefore, even if the skull has a high heritability component compared to the postcranial skeleton, its size and shape are plastic as well. This makes it possible to deform under particular circumstances, such as with developmentally acquired or genetic anomalies (Barners 2012); pathological conditions (Ortner 2003); childbirth; pre- and post-natal positional conditions (Graham, Sanchez-Lara 2015), and cultural, ritual or occupational practices (Duncan, Hofling 2011; Tiesler 2014).

Among different cultures and in different historical periods, humans have decorated their bodies for many reasons, creating traditions of dress, headgear, jewellery, henna, hairstyles, head shaving and face painting (Sørensen 1997; Choi 2006; Della Casa, Witt 2010). Body decoration may be achieved not only by the application of accessories or ephemeral ornamentations, but may result from forms of body shaping or even mutilation. In some instances, temporary body decoration is not considered sufficient for the prescribed aesthetic purposes, and in these cases, forms of body manipulation causing permanent changes in the physiognomy of an individual are preferred (Gerszten, Gerszten 1995). These include scarring; circumcision; foot binding, and chest, neck, tooth or head shaping (Flower 1881). The sphere of reasons for and meanings of body shaping is more complicated than individual aesthetic choice. In fact, the body as a means of nonverbal communication may transcend the individual sphere, rather reflecting more complex meanings that concern and reflect instead a system of beliefs developed and moulded by the group or the society to which the individual belongs. Investigations on forms of body decoration have revealed 'an interplay between individual display, group norms of attire and social context' (Johnson 2001: 418). Because of its widespread and long-lasting character, one of the most investigated forms of permanent body modification is headshaping, or artificial cranial deformation.

Headshaping: definition and classification

The term 'headshaping' has been used to refer to the modification of the shape of the head for social and cultural reasons. Thus, because of its aetiology, it can be distinguished from both natural deformations and malformations (Lorentz 2009). Headshaping has been defined as the dynamic distortion of the normal vectors of neurocranial growth with the application of external force (Moss 1958: 275). Tiesler (2014: 34) described this process as a 'dynamic interaction between culturally induced extrinsic mechanical stresses and intrinsic compensatory forces, which redirect head expansion'. Mechanical external tensions redirect the intrinsic growth vectors toward unrestricted cranial regions (Moss 1958; Tiesler 2014). Beyond the directly and indirectly affected areas of the neurocranium, secondary changes due to compensatory forces may be observed in the basicranium and splanchnocranium as well (Antón 1989). The headshaping process depends on various factors such as circumstances (Lorentz 2009), as well as techniques and applied devices (Imbelloni 1925). Circumstances define its intentionality. In many cases, although the purpose of modification is not primary, such as in therapeutic treatments (Hasluck 1947; Erdal 2013) and procedures promoting the development of infant motor skills (Hudson 1966), the resulting effects are aesthetically appreciated and thus well accepted and desired (Dingwall 1931; Lorentz 2009).

Intentional headshaping is performed immediately after birth by infant caretakers such as mothers or midwives, for a duration of about six months to one year, or even more (Dingwall 1931). The duration of the moulding process may depend on several factors such as the type of shaping device, the technique, the physical and behavioural development of the baby, and cultural and traditional components. In some cases, or with particular techniques or at particular ages, the results may be permanent. According to Tiesler (2014), intentional headshaping is practicable before the third year of age of an individual, when the active cerebral expansion which stimulates cranial growth adjustment is almost complete. Intentional headshaping may be achieved with manual moulding or cradling, or with the application of shaping devices such as bandages, headdresses, boards, stones, barks, pads and strings (Dingwall 1931). Differences in morphology and/or degrees of severity between types, or within the same type, depend on several factors. These include the type or combination of deforming apparatuses; the size and shape; the flexibility and thickness, and the area, orientation, amount, nature and duration of the applied pressure. Other factors include the age of the individual at the beginning of the headshaping process, the reaction of the bone tissue to the external mechanical pressure, malnutrition (Imbelloni 1925; Özbek 1974a; 2001; Aufderheide et al. 1998; Tiesler 2012), and the presence of congenital or genetic diseases or anomalies that can accelerate or hinder the shaping process (Ortner 2003). The most common intentional headshaping types are tabular and circular, classified into oblique and vertical directionality (Imbelloni 1925; Moss 1958; Tiesler 2014).

Tabular headshaping may result from the application of free boards or stones fastened by bands, leather straps or strings, or most commonly from the prolonged dorsal decubitus of the baby on a flat surface, such as cradleboards or cradles (Imbelloni 1925; Dingwall 1931). According to the employed techniques and devices, mechanical forces may unidirectionally affect only the frontal or the posterior portion of the skull, or both. Generally, posterior and anterior-posterior tabular types result in a brachycranic and hypsicranic shape of the skull due to its anteroposterior shortening, mediolateral widening and superoinferior development. More acute parietal angles, parietal tubera bulging and asymmetry (plagiocephaly) are diagnostic features of this headshaping type as well (Antón 1989; Kohn et al. 1995; Pomeroy et al. 2009).

Circular headshaping occurs when one or more bandages, strings, tree barks and strings, handkerchiefs or headgears envelop the head, generally from the frontal to the occipital or lambda region, passing over the parieto-temporal bones (Dingwall 1931). In some instances, a secondary device may cross the region behind the coronal suture going down to the lower part of the mandible (Lambert 1979; Özbek 2001). Compared to tabular headshaping, the multi-directionality of the applied forces provides a more symmetrical moulding. In general, flattening or depression on the frontal, the occipital or lambda region, postcoronal depression, parietal grooving and parietal tubera bulging are the most common diagnostic morphological alterations of this headshaping type (Erdal, D'Amico 2021). Mediolateral development of the skull is prevented, while its anteroposterior growth is allowed. Oblique circular headshaping is characterised by a postero- and/or superior development in length and a decrease in width. The skull acquires an ovoid shape *in norma lateralis* and *superioris*. Conversely, in the vertical subtype, height develops more than length so that the skull appears more conical in shape (Kohn et al. 1993; Pomeroy et al. 2009).

In Anatolia, the phenomenon of artificial headshaping is known from a small number of case-specific studies (Eroğlu 2015; 2016), or by the archaeological and anthropological information related to the investigations at sites where shaped skulls were found (Şenyürek, Tunakan 1951; Mellink, Angel 1968; Angel 1976; Özbek 1984; 2001; Alpagut 1986; Duyar, Atamtürk 2010; Miyake 2010; Erdal 2013; 2019; Aytek et al. 2020; 2021). Nevertheless, an overall assessment and interpretation of all the known examples does not exist yet. For that reason, the main aim of this study is to provide a diachronic representation of the distribution of headshaping in Anatolia, as well as an explanation of this practice from prehistory to the present day by evaluating the effects of socio-economic, political and cultural transformations – as well as the population movements that characterise different chronological periods and geographical contexts beyond Anatolia – on its types, frequency, severity and meanings. To that end, archaeological, anthropological, ethnographic and genetic data have been examined.

Intentional headshaping in Anatolia

Intentional headshaping in Anatolia is known to have been a common practice from at least the Neolithic period, and is still performed in the present day in some traditional communities (Erdal, D'Amico 2020). The presence of different types has remained a controversial topic. On one hand, the high frequency of circularly shaped skulls might be related to the prominence of their main morphological features, whilst the limited information on tabular headshaping has been influenced by the general anthropological consensus that, for a long time, has accounted for this cranial morphology as normal and even racially determined (Erdal, D'Amico 2020). New investigations (conducted by the author of this chapter) have revealed the existence of another type, defined as 'composite', which combines morphological cranial alterations observed both in circular and tabular headshaping. Based on these features, its development has been explained through the application of both textiles and the dorsal decubitus of the baby in a cradle (Fig. 6.1).

In this study, diachronic analysis of the typological distribution of headshaping in Anatolia has been done based on already published works, and by using new data from investigations conducted by the author.

Circular headshaping, performed by one or especially two bandages, is the oldest representative type. The circularly shaped skull of a child characterised by a post-coronal depression was found at the site of Salat Cami Yanı in Diyarbakır province, and was dated to the Pottery Neolithic period (Miyake 2010) (Fig. 6.2, Tab. 6.1). An

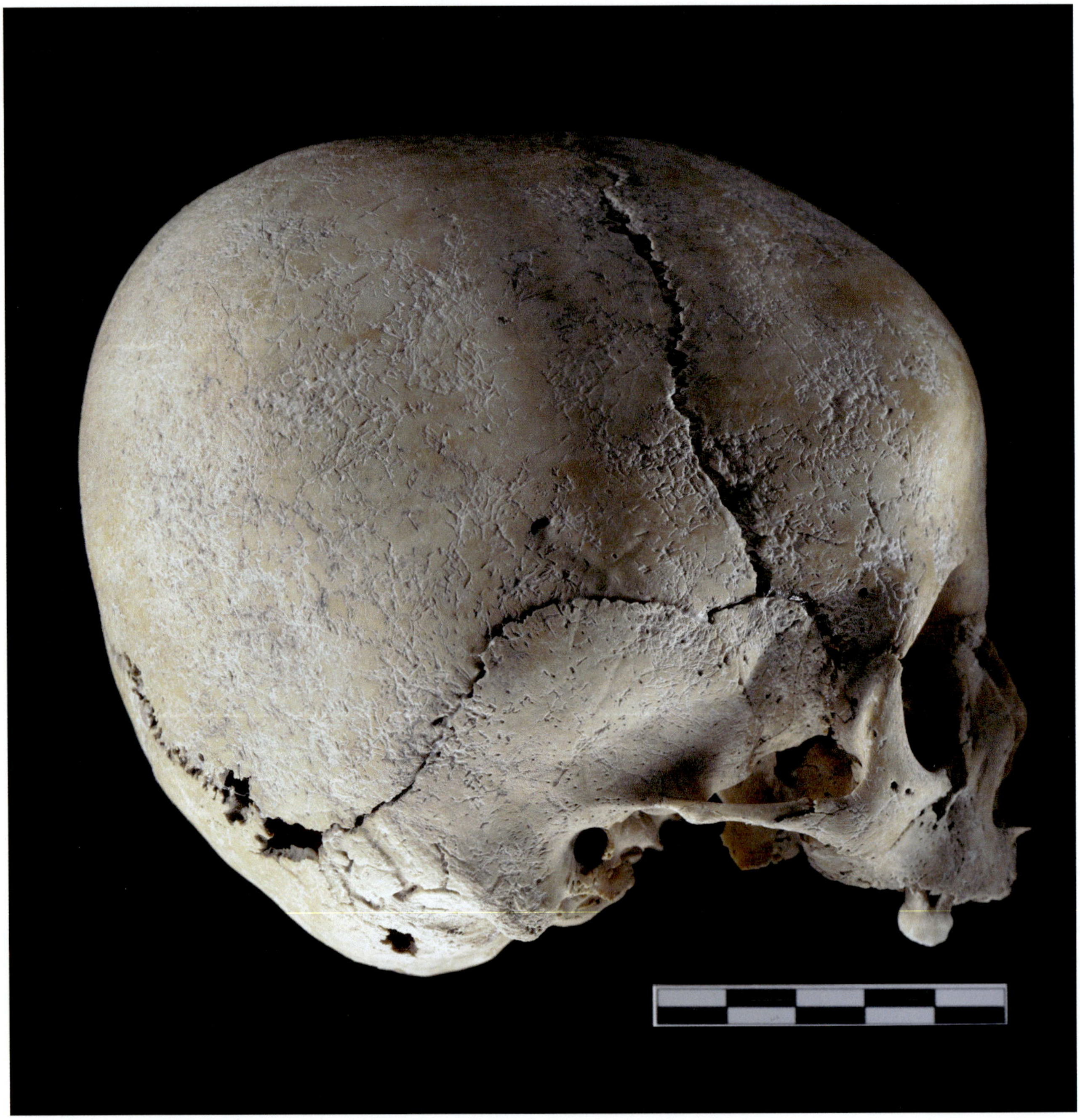

Fig. 6.1. Composite headshaping in a subadult from Arslantepe (photo by author).

Site	*Chronology*	*Headshaping type*	*References*
Salat Cami Yanı	Pottery Neolithic	Circular	Miyake 2010
Hakemi Use	Late Neolithic	Circular	Erdal 2013
Kurban Höyük	Late Neolithic	Circular	Alpagut 1986
Değirmentepe	Chalcolithic	Circular	Özbek 2001
Şeyh Höyük	Chalcolithic	Circular	Şenyürek, Tunakan 1951
Havuz Mevkii	Chalcolithic	Circular	Eroğlu 2015
Çadır Höyük	Late Chalcolithic	Circular	Erdal 2019a; Erdal, D'Amico 2021
Arslantepe	Late Chalcolithic	Circular	Erdal 2019a; Erdal, D'Amico 2020, 2021
İkiztepe	Late Chalcolithic / Early Bronze Age	Circular Plano-occipital flattening	Unpublished data
Karataş	Early Bronze Age	Plano-occipital flattening Circular (Cypriot?)	Mellink, Angel 1968; Angel 1976
Resuloğlu	Early Bronze Age	Circular (Cypriot?)	Duyar, Atamtürk 2010
Hayaz Höyük	Middle Bronze Age	Circular	Özbek 1984
Bakla Tepe	Bronze Age	Circular	Erdal, personal communication, March 2021
Zeviya Tivilki	Iron Age	Circular	Eroğlu 2016
Gordion	Iron Age	Circular	Unpublished data
Myndos	4th C. BC – AD 4th C.	Circular	Aytek et al. 2021
Tefenni	Early Byzantine	Circular (composite?)	Aytek et al. 2020
Muğla	Unknown	Circular	Unpublished data
Arslantepe-Malatya	Byzantine-medieval	Plano-occipital flattening Circular Composite	Erdal, D'Amico 2020
İznik-Bursa	Byzantine-medieval	Plano-occipital flattening Circular Composite	Unpublished data
Oluz Höyük-Amasya	Byzantine-medieval	Plano-occipital flattening Circular Composite	Unpublished data
Amasya	Modern	Plano-occipital flattening	Erdal, D'Amico 2020
Tokat	Modern	Plano-occipital flattening	Erdal, D'Amico 2020
Elazığ	Modern	Plano-occipital flattening	Erdal, D'Amico 2020
Sinop	Modern	Plano-occipital flattening	Erdal, D'Amico 2020
Malatya	Modern	Plano-occipital flattening	Erdal, D'Amico 2020
Diyarbakır	Modern	Circular	Erdal 2013; Erdal, D'Amico 2020
Gaziantep	Modern	Circular	Erdal, D'Amico 2020
Karataş	Modern	Plano-occipital flattening	Mellink, Angel 1968
Yürüks of south Anatolia	Modern	Circular	Lorentz 2010
Ansaris of Antioch	Modern	Plano-occipital flattening Circular	Chantre 1886

Tab. 6.1. Chronological and typological distribution of headshaping in Anatolia

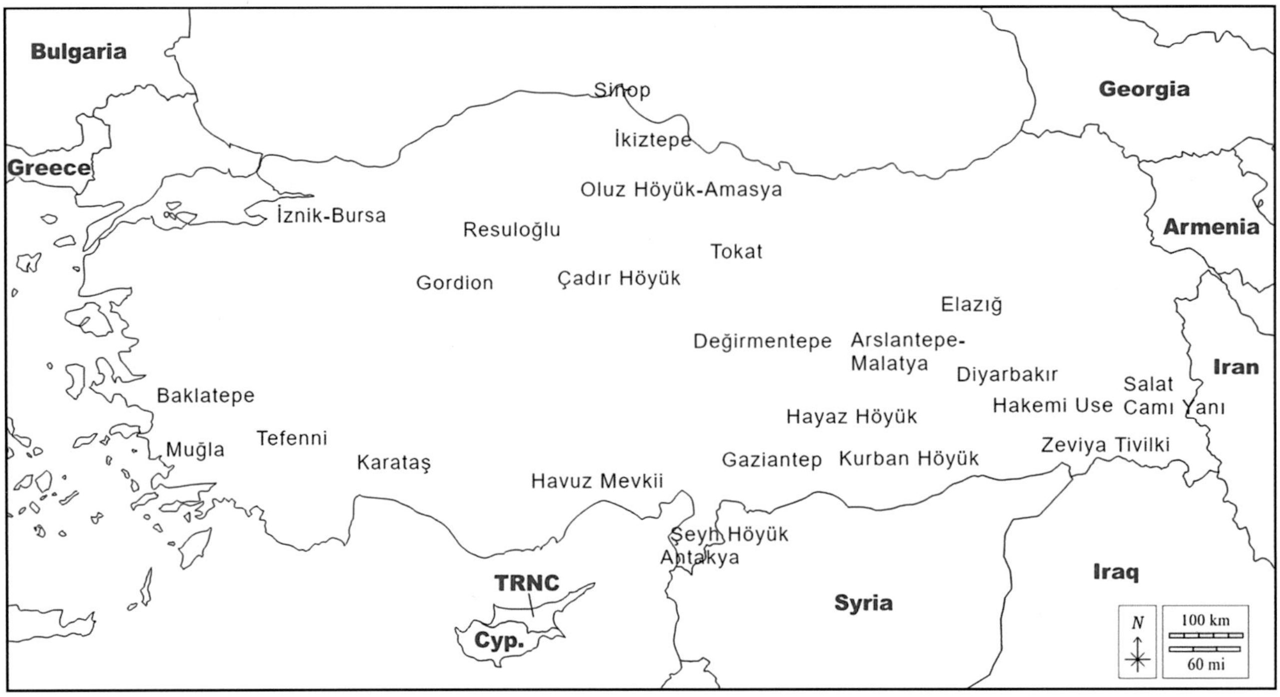

Fig. 6.2. Geographical distribution of intentional headshaping in Anatolia from the Neolithic to the present day (map by author).

adult female with a post-coronal depression was discovered at the Late Neolithic site of Kurban Höyük (Alpagut 1986) in the province of Sanlıurfa (Fig. 6.2, Tab. 6.1). From the same chronological period are the human skeletal remains found at Hakemi Use, in Diyarbakir province, where there was a very high frequency of circular headshaping (Erdal 2013) (Fig. 6.2, Tab. 6.1). The skulls of individuals from different age groups and both sexes showed signs of a double bandage reaching from the frontal to the occipital region, and from the coronal/post-coronal area to the lower jaw (Erdal 2013).

Circular headshaping continues to be observed during the Chalcolithic Period as well. Known examples come from the southeastern sites of Değirmentepe (Özbek 2001), Şeyh Höyük (Şenyürek, Tunakan 1951) and Havuz Mevkii (Eroğlu 2015), from Arslantepe in eastern Anatolia and Çadır Höyük (Erdal 2019a; Frangipane et al. 2019) in the central Anatolian region (Fig. 6.2, Tab. 6.1). Skulls from Değirmentepe, all belonging to subadults, were shaped using both one and two bandages. In particular, it has been suggested that younger individuals were wrapped with a single bandage and older ones with a double bandage (Özbek 2001). Although morphological cranial variations are very similar to those observed in other Anatolian examples dated to the Neolithic period, depressions are the most prominent so far observed, in some cases reaching nearly 10mm in depth. Shaped crania from Şeyh Höyük, belonging to both adult females (more prominent) and males, showed posterosuperior elongation and signs of both a single and a double bandage (Şenyürek, Tunakan 1951). Frontal and post-coronal depressions were also observed on the skull of a juvenile found at Havuz Mevkii (Eroğlu 2015), as well as on a child at Çadır Höyük and on an infant at Arslantepe (Erdal 2019a; Frangipane et al. 2019; Erdal, D'Amico 2021). At Arslantepe, a very long and narrow skull of an adult female, most probably shaped with the application of a bandage enveloping the entire head with the same amount of pressure or by a sort of headgear, was also discovered.

New analysis conducted on the cranial remains from İkiztepe, in the Black Sea region (Fig. 6.2), and dated to the Chalcolithic-Early Bronze Age, has identified the presence of both circular and tabular headshaping (plano-occipital flattening) (Tab. 6.1). The circular type is higher in frequency but less severe than those of the Late Neolithic and Early Chalcolithic periods; the tabular type appears as an episodic phenomenon occurring no earlier than the Early Bronze Age. Beyond İkiztepe, the oldest Anatolian examples of this type are known from the southwestern site of Karataş (Mellink, Angel 1968; Angel 1976) (Fig. 6.2, Tab. 6.1). Even after the appearance of tabular headshaping, the circular type continued to be widespread, as proven by the examples found at Karataş (Mellink, Angel 1968; Angel 1976), Hayaz Höyük (Özbek 1984) in the southeast, Resuloğlu in northern Anatolia and Bakla Tepe in the district of Menderes near Izmir (Fig. 6.2, Tab. 6.1). At Resuloğlu, morphological cranial alterations caused by the application of a single bandage enveloping the head from the post-coronal region down to the occipital area, were recognised on the skulls of four subadults (Duyar, Atamtürk 2010). A similar form of headshaping was also observed at Karataş by Angel and linked to the Cypriot Type common on the island of Cyprus during the Late Neolithic Period (Angel 1976). At Bakla Tepe, pre-coronal and post-coronal depressions were observed on the skull of an infant (pers. comm. Y.S. Erdal, March 2021).

During the Iron Age, examples of severe circular headshaping are known from the sites of Gordion in central Anatolia and Zeviya Tivilki in the southeast (Eroğlu 2016) (Fig. 6.2, Tab. 6.1). The examples from Gordion belong to an adult female and a subadult, and are characterised by very evident posterosuperior neurocranial development, although the subadult, similar to the shaped adult female skull found at Arslantepe, does not show any sign of depression typical of circular headshaping. From Zeviya Tivilki, the skull of an adult male showed a posterosuperior neurocranial development and depressions in the frontal and post-coronal regions. No case of tabular headshaping is yet known for this chronological period.

It is only at the beginning of the Byzantine-medieval period that tabular headshaping reappears, coexisting with composite and circular types; this latter type is performed with the application of both one and two bandages. Compared to the previous period, circular headshaping in the medieval period is characterised by lower frequencies and a slight severity of development. Unfortunately, very few studies on intentional headshaping in Anatolia exist for this chronological period. A published case of circular modification is known from the Necropolis of Myndos (fourth century BC to AD fourth century), where a post-coronal depression was observed on the skull of an adult (Aytek et al. 2021). Similarly, at the site of Tefenni (Aytek et al. 2020) (Fig. 6.2), a post-coronal depression was recognised on the skull of a child dated to the Early Byzantine period (Aytek et al. 2020) (Tab. 6.1). Based on the published photograph (Aytek et al. 2020), the presence of plano-occipital flattening may also be suggested (Tab. 6.1). At Arslantepe, İznik and Oluz Höyük, all three headshaping types have been observed. However, the frequency of circular headshaping is very low and its severity slighter (Fig. 6.1). The composite type is very common at Arslantepe while the tabular is the most common at İznik and Oluz Höyük. Both the types show evident degrees of severity.

In the last two centuries, while the composite type is no longer observed, circular and tabular headshaping continues to be practised in different regions of the Anatolian peninsula. The tabular type is performed by people living in the northeastern regions of Malatya, Elazığ, Amasya, Tokat and Sinop (Erdal 2013; Erdal, D'Amico 2020) and by the inhabitants of the southwestern city of Semayük (Mellink, Angel 1968) (Fig. 6.2, Tab. 6.1), while circular headshaping is practised in the southeastern cities of Diyarbakır and Gaziantep (Erdal, D'Amico 2020), and by the nomadic Yürüks who mainly live in Mersin province and in the central-eastern regions of Türkiye (Lorentz 2010) (Fig. 6.2, Tab. 6.1). Chantre (1886) reported the presence of both circular and tabular headshaping among the Ansaris living in the southeastern city of Antakya as well (Fig. 6.2, Tab. 6.1).

Reasons behind intentional headshaping in Anatolia

Bibliographic and ethnographic investigations have shown that the main motives behind this cultural phenomenon concern both the individual and the social sphere, such as the construction of gender, status and ethnic differences. It must be highlighted, though, that intentional headshaping is in premise accepted as aesthetically appealing. According to Erdal (2013), almost all members of the Neolithic settlement of Hakemi Use had intentionally modified skulls, and this was most probably for aesthetic purposes. Some traditional Anatolian people also value intentionally elongated heads (e.g. in Diyarbakır and Gaziantep) and/or plano occipital flattening (in Sinop, Amasya, Tokat, Ördüzü and Malatya) as appealing (Erdal, D'Amico 2020). Beyond aesthetic reasons, therapeutic treatments and procedures aimed at enhancing intelligence have been proposed as other possible explanations (FitzSimmons et al. 1998; Erdal 2013). Contemporary populations of Türkiye, such as Arabs, Kurds and Turkmens, practise intentional headshaping to protect cranial fontanelles and correct ear deformities (Erdal 2013). Özbek (2001) suggested that intentional headshaping was exclusively reserved for women at the site of Değirmentepe during the Chalcolithic period. Ethnographic research conducted at Ordüzü (Erdal, D'Amico 2020) revealed that headshaping achieved by cradling is intentionally practised in order to differentiate Kurds living in this area from those living in neighbouring territories, who are characterised by long heads. Examples of intentional headshaping from Çadır Höyük and Arslantepe, and dated to the Late Chalcolithic

Period, have been evaluated with regard to the construction and enhancement of hereditarily transmitted forms of social status (Erdal, D'Amico 2021). Since artificial headshaping may reflect different aspects of socio-cultural life tightly embedded into one another, all the meanings attributed to this cultural practice are consequently intertwined so that none of them can be reductively considered and interpreted individually.

Beyond Anatolia

Intentional headshaping is not isolated and circumscribed to the Anatolian peninsula, rather it has been very common across much more of the Near East both in ancient times and in the present. Its geographical distribution extends from the Levantine region and Cyprus to the northeastern territories of Iraq and the western regions of Iran, and westwards to Greece. Indeed, while still debated, the Neanderthal cranial remains discovered at Shanidar Cave in Iraq, dating to the Middle Palaeolithic period (Trinkaus 1982; Chech et al. 1999), have generally been accepted as the oldest known examples of intentional headshaping in this geographical region. In any case, headshaping is observed most frequently in the earlier phases of the (pre-ceramic) Neolithic period.

Circular headshaping was the most common type in the Neolithic for the whole of the Near East. Examples are known from the Levantine sites of Jericho (PPNA-B) (Kurth, Röhrer 1981), Tel Roim West (Eshed, Nadel 2015) and Bouqras (Meiklejohn et al. 1992), from the Iraqi site of Shanidar Cave, and from the western Iranian sites of Tepe Ghenil (Meiklejohn et al. 1992), Choga Sefid (Hole 1977), Ganj Dareh Tepe (Lambert 1979), Tepe Abdul Hosein (Lorentz 2017) and Ali Kosh (Lorentz 2010). This widespread distribution continued during the Chalcolithic Period with examples discovered across all of Mesopotamia. Moulded skulls were found at Byblos in Lebanon (Özbek 1974a; 1974b), Tell Zeidan in Syria (Stein 2011), Eridu (Kiszely 1978) and Tell Madhur (Downs 1984) in Iraq, and at the Iranian sites of Choga Mish (Ortner 1996), Seh Gabi (Lambert 1979) and Qumrud (Lorentz 2010). Circularly shaped crania from the Iraqi site of Tell Arpachiyah (Molleson, Campbell 1995) belong to both the Halaf and Ubaid cultures.

The development of interregional exchange networks of goods and raw sources from the Neolithic period onwards is well known. It has even been suggested that some of these networks were likely established much earlier than the Neolithic (Bressy et al. 2005; Kılınç al. 2017). The existence of well organised interregional trade networks between southeastern-central Anatolia, northeastern Syria, northern Iraq and northwestern Iran during the Halaf period is proven by patterns of domestic architecture (Campbell et al. 1999; Akkermans 2000), and by the finding of many artefacts, as well as obsidian and other precious stones that would have been brought from central and southeastern Anatolia (Bressy et al. 2005; Kılınç et al. 2016). In this clear system of connectivity, the existence of exchanges occurring at a deeper social level regarding ideas, beliefs and customs may easily be argued for. Based on settlement patterns, the lack of religious or ceremonial buildings and the absence of social status distinctions in burial customs, the Halafian socio-economic structure has been acknowledged as egalitarian and based on kinship groups or clans (Frangipane 2007; 2015). According to Frangipane (2007: 161), similarities in the material culture observed among all the Halaf communities, even those furthest from each other, have to be explained in terms of a need for economic and social cohesion. Genomic analyses attest to a mixed ancestry on a genetic cline formed after an admixture event occurred in at least the sixth millennium BC between peoples of western Anatolia, the Zagros territories and even the southern Caucasus (Skourtanioti et al. 2020). With such a socio-economic context and genetic makeup, the wide distribution of a unique headshaping type (circular) from the Zagros mountains of Iran to the Anatolian mountains of the Taurus and beyond might be explained as part of the diffusion of Hassuna, Samarra, Halaf (Akkermans 2000) and Ubaid (Carter, Philip 2010) cultures, and may be interpreted in terms of visually constructed forms of social affiliation within egalitarian societies attached to the same cultural sphere.

By the fifth millennium BC, cultural elements of southern Mesopotamia were spreading northeastwards. Settlement patterns and pottery reflect this cultural flow (Carter, Philip 2010; Parker 2010). An increase in technological and morphological standardisation of pottery, as well as a development in the size of regional centres and monumental architecture, may also be interpreted within the framework of the socio-economic changes occurring during this period (Greenberg, Palumbi 2014). Clay sealings, seals and tokens, already in use during the Halaf period (Campbell et al. 1999; Akkermans 2000), are now discovered in temples and houses (Frangipane 2007). All this data not only suggests a system of economic networks between regions far away from each other, but also the development of hierarchically organised societies characterised by competition and differentiation of resource access (Frangipane 2007; 2015).

In this chronological period, changes in intentional headshaping may be identified at a semantic level rather than a typological one. According to Lorentz (2010: 140–

41), circumferential headshaping during the Ubaid period may be related to the emergence of social complexity and to the need for differentiation. As proposed by Molleson et al. (1995: 50–51), at the site of Tell Arpachiyah in Iraq, intentional headshaping was practised on members of a hereditary group or class – priests, potters or princes. The emergence of political institutions and elites that coordinated collective labour and controlled resource access and distribution during the Late Chalcolithic presaged the formation of centralised early state polities, which was a reflection of cultural influences from southern Mesopotamia (Greenberg, Palumbi 2014). Although no typological differences have been recognised compared to the previous periods, a decrease in frequency might suggest a connection between intentional headshaping and special status among particular community members, most probably high social rank. In this regard, the examples of Arslantepe and Çadır Höyük in Anatolia (Fig. 6.2, Tab. 6.1), dated to the Late Chalcolithic period, are distinctive (Erdal, D'Amico 2021). In these two archaeological contexts, the very low frequency of intentional headshaping, the socio-political relevance of the funerary contexts where shaped individuals were discovered and the presence of precious grave goods fit well with the aforementioned socio-economic and political framework. In light of these scenarios, intentional headshaping in Anatolia and around the Near East might be interpreted as a visual and symbolic attempt to enhance hereditary transmitted forms of social status within complex, hierarchically organised societies.

At the end of the fourth and the beginning of the third millennium BC, while cultural standardisation was occurring over a very wide area from the Persian Gulf to southeastern Anatolia, there was also an increase in northern influences from the Caucasus. The intensification of sheep husbandry and metallurgy, the emergence of totally burnished Red-Black Ware, new building techniques and burial customs have provided some clues about these new cultural flows (Frangipane et al. 2001; Greenberg, Palumbi 2014). Arslantepe and İkiztepe show traces of material culture strongly connected to this northern culture. Nevertheless, the results of genomic analyses have attested a genetic continuity in the populations occupying territories of the western Anatolia, southern Caucasus and Zagros regions during the Late Chalcolithic and/or Early Bronze Age (Skourtanioti et al. 2020). It has been suggested that the pastoralists occupying the site of Arslantepe during the Early Bronze Age were most probably well-established local people rather than intrusive groups from the Caucasus (Skourtanioti et al. 2020). This conclusion refutes the previously held hypothesis of an invasion which was proposed to explain the socio-cultural transformations that had occurred in Anatolia with the advent of the Bronze Age (Şenyürek 1941). As well as in the material culture, a variation in intentional headshaping type was observed at the beginning of the Bronze Age. Tabular headshaping, and in particular plano-occipital flattening – previously observed in a very circumscribed geographical area including the Near Eastern Aceramic Neolithic sites of Jericho and Cyprus (Kurth, Röhrer 1981) – spanned many other areas of the Near East. Examples are known from the Levantine sites of Ras Shamra and Tepe Sialk in Iran (Soto-Heim 1986), and from several Cypriot sites such as Sotira, Hala Sultan Tekke, Enkomi, Marki, Melia, Vounous and Bamboula (Angel 1936; Bolger 2003; Lorentz 2008). It also made its first appearance in Anatolia at the sites of Karataş and İkiztepe (Mellink, Angel 1968; Angel 1976). Nevertheless, it must be noted that in Anatolia, the circular type continued to be the most widespread until the Iron Age, while the tabular type was only episodic.

Based on genomic analyses, it is possible to state that the occurrence of this new headshaping type in Anatolia was not related to a foreign presence from the Caucasus. In light of this evidence, it is interesting to note similarities in the material culture, architecture and funerary customs between Anatolia, Cyprus, Crete and the southern Levant during the third millennium BC (Knapp 2008). Technological, morphological and decorative similarities have been recognised in Early Bronze Age vessels from Tarsus (Cilicia) and Karataş-Semayük (Lycia) in Anatolia and Cyprus (from the Philia phase). Other affinities have been identified between spindle whorls belonging to the Philia phase and to the Cilician Early Bronze Age, as well as between clay hobs found at Marki Alonia and those discovered in Anatolia and the southern Levant (Knapp 2008). An interaction between Anatolia and Cyprus has been proposed with regard to the procurement of metal sources and artefacts from the island (Knapp 2008). Thus, the presence of the tabular type in Anatolia might be explained by considering the long-term economic and cultural interactions with Cyprus. This suggestion may also be supported by Mellink and Angel's (1967) hypothesised connection between the Karataş and Cyprus people on the basis of the similarities in their morphological cranial features. According to Angel (1976), occipital flattening might have been brought from Cyprus, where the type was known from the Early Neolithic period. Moreover, the so-called Post-bregmatic or Cypriot type identified on Cyprus and dated to the Late Cypriot period (Bolger 2003) might resemble that observed at the Anatolian sites of Karataş (Angel 1976) and Resuloğlu (Duyar, Atamtürk 2010).

According to the results of craniological studies conducted on ancient Anatolian populations, a homogeneous Anatolian genetic structure existed until the second millennium BC. Despite the episodic appearance of tabular headshaping in Anatolia, the higher frequency of circular headshaping – examples of which are known from the Bronze Age site of Bakla Tepe and the Iron Age sites of Gordion and Zeviya Tivilki (Eroğlu 2016) (Fig. 6.2, Tab. 6.1) – could support the hypothesis of a continuing and persistent genetic homogeneity. However, it seems that after the Iron Age, the genetic makeup of the Anatolians was subjected to alterations (see also Eroğlu, this volume). Invasions, migrations and wars have been suggested as the most important factors behind these changes. The increase in brachycephalism, for example, was explained by the arrival of different ethnic groups (Şenyürek 1941; Saatçioğlu 1982). The hypothesis of the arrival of new people should be interpreted not only by taking into consideration the process of Aegean colonisation but also through the persistence of contact between Anatolians, Phoenicians and Cypriots. Effects of these interactions are witnessed in Cyprus's Cyclopean fortifications; in the dog-leg gates resembling the constructions found at Boğazköy in Anatolia, Mycenae and Tiryns in Greece, and Shechem in the southern Levant; in the large rooms with a central hearth, of which evidence has been found at Cyprus, Tarsus and at several Philistine sites in the southern Levant; as well as in the widespread distribution of Mycenaean pottery (Knapp 2008).

It is only from the Byzantine-medieval period that tabular headshaping spread in the Near East and Anatolia, where, together with the composite type, it is observed with high frequency and degrees of severity. The circular type, on the other hand, is rarely observed, and with slight developmental severity compared to the previous chronological periods. Together with the Anatolian sites of Arslantepe (Malatya), İznik (Bursa), Oluz Höyük (Amasya) (Fig. 6.2, Tab. 6.1), some examples of tabular headshaping dated to the medieval period are known from the Hellenistic-Byzantine Israeli sites of En Gedi, Yavne Yam and Jericho (Arensburg, Hershkovitz 1988), and from the Islamic-period site of Bolghasian in Iran (Soto-Heim 1986). This scenario continues until more recent times, when the composite type is no longer observed, while circular and tabular headshaping survive in different regions of the Anatolian peninsula (Fig. 6.2, Tab. 6.1).

Biological distance and aDNA studies, as well as historical events, provide important clues to better understand and interpret the distribution of intentional headshaping in the medieval and post-medieval periods. In the 11th century, a new population, the Seljuk Turks (Mongoloid), spread westwards into Anatolia. This migration, together with those of Tatars and groups from the Caucasus, continued until the 18th and 19th centuries (Eroğlu, Erdal 2008). Genetic investigations (Di Benedetto 2001) confirmed the existence of an Asiatic gene flow during this period. Biological distance data on shovel-shaped teeth, a non-metric trait typical of Mongoloids, was found to have a high frequency and severe degree of development in Anatolia, especially during the Byzantine and Ottoman periods. Moreover, its average frequency in Anatolia was closer to that found in Asian populations than European ones (Eroğlu 2012). In the early 20th century, a variation in gene flow occurred. Immigrations no longer came from the East, but from Balkan states (Bulgaria, Greece, Romania, Yugoslavia), especially after the establishment of the Republic of Türkiye (Eroğlu 2016). Some studies on *foramen supraorbitalis* and *canalis hypoglossi* (Eroğlu 2010; 2016), which are most frequently observed among west Eurasian and ancient African communities compared to Asian populations, had the highest frequencies in Anatolia during the Byzantine period and in the 20th century. According to DNA analyses conducted by Cinnoğlu et al. (2004), 94.1% of the Turkish Y chromosome is shared with European and Middle Eastern populations, while 3.4% with Asian populations, 1.5% with Indian populations and 1.2% with African populations. A study on nine Alu intermediate polymorphisms (Şekeryapan 2005) demonstrated that there is a close genetic relation between the Anatolian population and those of the Balkans, the Caucasus and the Middle East, compared to Asian populations. Studies on the mylohyoid canal (Eroğlu 2011) and *sutura metopica* (Eroğlu 2008) revealed frequencies that are in between European and Asian populations, thus reflecting the heterogeneity of the Anatolian population. In such a genetic framework, different types of intentional headshaping might have been used as a means of ethnic differentiation.

Conclusion

Intentional headshaping is a deeply established cultural practice performed from the Neolithic to the present day in Anatolia and around the Near East. Typological and semantic variations have been observed during different chronological periods, most probably as an effect of socio-economic and political changes, as well as during shifts in the genetic composition. From the Neolithic to the Late Chalcolithic and/or Early Bronze Age, the exclusive spread of intentional circular headshaping in Anatolia, as well as in the southern Caucasus and the Zagros regions, represents an attempt to construct an ethnic membership reflecting not only the genetic homogeneity existing between people living in these geographical areas, but also (and especially) related to the diffusion of the Halaf and Ubaid cultures. The intensive interregional trade networks were at the base of the diffusion of ideas, beliefs and cultural practices including intentional headshaping. Within such a socio-economic framework, this cultural phenomenon represented a means by which to establish and strengthen forms of social affiliation in terms of ethnic belonging within societies which were likely characterised by an egalitarian structure. With the beginning of hierarchical forms of socio-economic and political organisation, although changes in the type and severity of headshaping were not observed, its decrease in frequency suggests a relation with special members of the community, most probably those belonging to the high social ranks. This suggestion may be supported by the precious grave goods and the special funerary contexts characterising some individuals with intentionally shaped skulls dated to this chronological period. As a form of body modification performed immediately after birth, its employment as a symbolic attempt to transmit hereditary forms of power may be proposed. With the advent of the Bronze Age, interregional cultural influences started to involve a larger geographical area, including the southern Caucasus (Greenberg, Palumbi 2014). Nevertheless, the results of genomic analyses (Skourtanioti et al. 2020) have not revealed any genetic shift compared to the previous periods. The very rare occurrence of plano-occipital flattening, as well as the continuity of the circular type until at least the Iron Age in Anatolia, could be read within the context of episodic and feeble demic interactions between these geographical regions. In this regard, it is interesting to keep in mind the relations between Anatolia and Cyprus, where tabular headshaping was very widespread since the Early Neolithic (Knapp 2008). From this perspective, as already proposed by other scholars (Mellink, Angel 1967; Angel 1976), a Cypriot origin for Anatolian plano-occipital flattening might be suggested. Following the Bronze Age, although the circular type of headshaping continues to be observed, its frequency and severity decreased in favour of other types. Their co-occurrence reflects the population admixtures which occurred after the second millennium BC. Population movements during the Seljuk and Ottoman periods drove new genetic components, this time of Asian origins, which continued until the 20th century. In the modern period, a new variation in gene flow has occurred, with immigrants coming from Europe rather than Central and East Asia (Cinnioğlu et al. 2003). Within this migration scenario, the coexistence of different headshaping types may reflect the need to establish forms of social/ethnic differentiation.

Thus, it can be concluded that comprehension of the mechanisms underlying the diffusion of intentional headshaping in Anatolia cannot prescind from the geographical contextualisation of the phenomenon within a larger space, including the whole Near East, and from the consideration of the socio-economic and political events that shaped the history of these regions. Anatolia has always been a strategic bridge between Asia and Europe, and intentional headshaping, as a social product, is also a reflection of its historical, geographical, socio-economic and political position.

Bibliography

Akkermans, P.M.M.G. 2000: 'Old and new perspectives on the origins of the Halaf culture' *Subartu* 7: 43–54

Alpagut, B. 1986: 'The human skeletal remains from Kurban Höyük (Urfa province)' *Anatolica* 13: 149–74

Angel J.L. 1936: 'The human remains from Khirokitia' in P. Dikaios (ed.), *Khirokitia: Final Report on the Excavation of a Neolithic Settlement in Cyprus on Behalf of the Department of Antiquities, 1936–1946 1*. Cyprus, Government of Cyprus: 416–30

— 1976: 'Excavations in the Elmalı area, Lycia, 1975. Appendix: Early Bronze Karatas people and their cemeteries' *American Journal of Archaeology* 80.4: 385–91

Anton, S.C. 1989: 'Intentional cranial vault deformation and induced changes of the cranial base and face' *American Journal of Physical Anthropology* 79.2: 253–67

Arensburg, B., Hershkovitz, I. 1988: 'Cranial deformation and trephination in the Middle East' *Bulletins et Mémoires de La Société d'anthropologie de Paris* 5.3: 139–50

Aytek, A.İ., Yavuz, A.Y., Kesici, S., Bağci, E.B. 2021: 'Myndos nekropol iskeletlerinin paleoantropolojik analizi ve bir savaşçı mezarı' *Antropoloji* 41: 54–65

Aytek, A.İ., Yavuz, A.Y., Özbey, O.G., Şahin, E. 2020: 'Tefenni Kurtarma kazısından çıkan iskeletlerin antropolojik analizi ve bir kafatası deformasyonu örneği' *Antropoloji* 39: 49–56

Barnes, E. 2012: *Atlas of Developmental Field Anomalies of the Human Skeleton: A Paleopathology Perspective*. Hoboken, John Wiley & Sons

Bolger D.R. 2003: *Gender in Ancient Cyprus: Narratives of Social Change on a Mediterranean Island* 6. Walnut Creek, Rowman Altamira

Bonogofsky, M. (ed.) 2011: *The Bioarchaeology of the Human Head: Decapitation, Decoration, and Deformation.* Gainesville, University Press of Florida

Bressy, C., Poupeau, G., Yener, K.A. 2005: 'Cultural interactions during the Ubaid and Halaf periods: Tell Kurdu (Amuq Valley, Turkey) obsidian sourcing' *Journal of Archaeological Science* 32.10: 1560–65

Campbell, S., Carter, E., Healey, E., Anderson, S., Kennedy, A., Whitcher, S. 1999: 'Emerging complexity on the Kahramanmaras Plain, Turkey: the Domuztepe Project, 1995–1997' *American Journal of Archaeology* 103.3: 395–418

Carter, R.A., Philip, G. 2010: 'Deconstructing the Ubaid' in R.A. Carter, G. Philip (eds) *Beyond the Ubaid Transformation and Integration in the Late Prehistoric Societies of the Middle East No. 63*. Chicago, Oriental Institute of the University of Chicago: 1–22

Chantre, E. 1886: 'Étude de crânes d'Ansariés' *Publications de la Société Linnéenne de Lyon* 5.1: 179–85

Chech, M., Groves, C.P., Thorne, A., Trinkaus, E. 1999: 'A new reconstruction of the Shanidar 5 cranium' *Paléorient* 25.2: 143–46

Cheverud, J.M., Kohn, L.A.P., Konigsberg, L.W., Leigh, S.R. 1992: 'Effects of fronto-occipital artificial cranial vault modification on the cranial base and face' *American Journal of Physical Anthropology* 88.3: 323–45

Cheverud, J.M., Midkiff, J.E. 1992: 'Effects of fronto-occipital cranial reshaping on mandibular form' *American Journal of Physical Anthropology* 87.2: 167–71

Choi, N.Y. 2006 'Symbolism of hairstyles in Korea and Japan' *Asian Folklore Studies* 65.1: 69–86

Cinnioğlu, C., King, R., Kivisild, T., Kalfoğlu, E., Atasoy, S., Cavalleri, G.L., Underhill, P.A. 2004: 'Excavating Y-chromosome haplotype strata in Anatolia' *Human Genetics* 114.2: 127–48

Della Casa, P., Witt, C. 2010: 'Tattoos and body modifications in Antiquity' *Proceedings of the Sessions at the EAA Annual Meetings in The Hague and Oslo* 11: 9–13

Di Benedetto, G., Ergüven, A., Stenico, M., Castrì, L., Bertorelle, G., Togan, I., Barbujani, G. 2001: 'DNA diversity and population admixture in Anatolia' *American Journal of Physical Anthropology* 115.2: 144–56

Dingwall, E.J. 1931: *Artificial Cranial Deformation. A Contribution to the Study of Ethnic Mutilation.* London, John Bale, Sons & Danielsson, Ltd.

Downs, D. 1984: 'The human skeletal remains' *Sumer* 43: 127

Duncan, W.N., Hofling, C.A. 2011: 'Why the head? Cranial modification as protection and ensoulment among the Maya' *Ancient Mesoamerica* 22.1: 199–210

Duyar, I., Atamtürk, D. 2010: 'Erken Tunç Çağında Orta Anadolu'da ölü gömme âdeti, sağlık yapısı ve yaşam biçimi: Resuloğlu örneği' *Çorum Kazı ve Araştırmalar Sempozyumu*: 30–41

Erdal, Y.S. 2013: 'Life and death at Hakemi Use' in O.P. Nieuwenhuyse, R. Bernbeck, P. Akkermans, J. Rogasch (eds) *Interpreting the Late Neolithic of Upper Mesopotamia*. Brepols: 213–23

— 2019a: 'Interpreting subadult burials and head shaping at Çadır Höyük' *Journal of Eastern Mediterranean Archaeology and Heritage Studies* 7.3: 379–85

Erdal, Y.S., D'Amico, V. 2020: 'Plano-occipital flattening: intentional or untintentional treatment?' in F. Balossi Restelli, A. Cardarelli, G.M. Di Nocera, L. Manzanilla, L. Mori, G. Palumbi, H. Pittman (eds) *Pathways through Arslantepe. Essays in Honour of Marcella Frangipane*. Rome, Sapienza Università di Roma: 561–77

— 2021: 'Modified bodies: an interpretation of social identity embedded into bones' in N. Laneri (ed.) *The Sacred Body. Material Religion in Antiquity (MaReA) Volume 1*. Oxford, Oxbow Books: 115–31

Eroğlu, S. 2008: 'The frequency of metopism in Anatolian populations dated from the Neolithic to the first quarter of the 20th century' *Clinical Anatomy* 21.6: 471–78

— 2011: 'The incidence of mylohyoid bridging in prehistoric and historic Anatolian populations' *Eurasian Journal of Anthropology* 2.1: 15–26

— 2012: 'Eski Anadolu toplumlarında kürek biçimli diş sıklığı' in E. Güleç, İ. Özer, M. Sagır, B.K. Özer (eds), *Kuruluş Yıldönümü Anı Kitabı: Biyolojik Antropoloji*. Ankara, Ankara Üniversitesi Dil ve Tarih-Coğrafya Fakültesi Yayınları: 163–81

— 2015: 'Antropolojik verilerin Ilısu Barajı İnşaat Sahası Kurturma Projesi II: Kalkolitik Çağ Mardin' in A.T. Ökse, N. Erdoğan (eds), *Ilısu Barajı İnşaat Sahası Kurtarma Projesi Ilı: Kalkolıtık Çağ*. Mardin, Mardin Müze Müdürlüğü: 515–34

— 2016: 'Supraorbital foramen and hypoglossal canal bridging in ancient/modern Anatolian populations: Implications for worldwide population distribution' *Mediterranean Archaeology and Archaeometry* 16.1: 169–83

Eroğlu, S., Erdal, Y.S. 2008: 'Why did the frequency of palatine torus increase in the ancient Anatolian populations?' *HOMO – Journal of Comparative Human Biology* 59.5: 365–82

Eshed, V., Nadel, D. 2015: 'Changes in burial customs from the Pre-Pottery to the Pottery Neolithic periods in the Levant: the case-study of Tel Roim West, Northern Israel' *Paléorient* 41.2: 115–31

FitzSimmons, E., Prost, J.H., Peniston, S. 1998: 'Infant head molding: a cultural practice' *Archives of Family Medicine* 7.1: 88

Flower, W.H. 1881: *Fashion in Deformity: As Illustrated in the Customs of Barbarous and Civilised Races*. London, Macmillan and Co.

Frangipane, M. 2007: 'Different types of egalitarian societies and the development of inequality in early Mesopotamia' *World Archaeology* 39.2: 151–76

— 2015: 'Different types of multiethnic societies and different patterns of development and change in the prehistoric Near East' *Proceedings of the National Academy of Sciences of the United States of America* 112.30: 9182–89

Frangipane, M., Di Nocera, G.M., Hauptmann, A., Morbidelli, P., Palmieri, A., Sadori, L., Schultz M., Schmidt-Schultz, T. 2001: 'New symbols of a new power in a "royal" tomb from 3000 BC Arslantepe, Malatya (Turkey)' *Paléorient* 27.2: 105–39

Gerszten, P.C., Gerszten, E. 1995: 'Intentional cranial deformation: a disappearing form of self-mutilation' *Neurosurgery* 37.3: 374–82

Graham Jr, J.M., Sanchez-Lara, P.A. 2015: *Smith's Recognizable Patterns of Human Deformation*. Philadelphia, Elsevier Health Sciences

Greenberg, R., Palumbi, G. 2014: 'Corridors and colonies: comparing fourth–third millennia BC interactions in southeast Anatolia and the Levant' *The Cambridge Prehistory of the Bronze and Iron Age Mediterranean*: 111–38

Hasluck, M. 1947: 'Head-deformation in the Near East' *Man: A Monthly Record of Anthropological Science* 47: 142–59

Hauser, G., De Stefano, G.F., Bastianini, A., Capecchi, V., Guidotti, A., Bertelli, L. 1989: *Epigenetic Variants of the Human Skull*. Stuttgart, Schweizerbart

Hole, F. 1977: *Studies in the Archaeological History of the Deh Luran Plain: The Excavation of Chagha Sefid* (Memories of the Museum of Anthropology 9). Ann Arbor, University of Michigan Press

Hudson, C. 1966: 'Isometric advantages of the cradle board: a hypothesis' *American Anthropologist* 68.2: 470–74

Imbelloni, J. 1925: *Sur un appareil de déformation du crâne des anciens Humahuacas*. Göteborg, Elanders Boktryckeri Aktiebolag

Johnson, R. 2001: 'The anthropological study of body decoration as art: collective representations and the somatization of affect' *Fashion Theory* 5.4: 417–34

Kiszely, I. 1978: *The Origins of Artificial Cranial Formation in Eurasia from the Sixth Millennium BC to the Seventh Century AD 50*. Oxford, British Archaeological Reports

Kılınç, G.M., Omrak, A., Özer, F., Günther, T., Büyükkarakaya, A.M., Bıçakçı, E., Baird, D., Dönertaş, H.M., Ghalichi, A., Yaka, R., Koptekin, D., Açan, S.C., Parvizi, P., Krzewinska, M., Daskalaki, E.A., Yüncü, E., Dağtaş, N.D., Fairbairn, F., Pearson, J., Mustafaoğlu, G., Erdal, Y.S., Çakan, Y.G., Togan, İ., Somel, M., Stora, J., Jakobsson, M., Götherström, A. 2016: 'The demographic development of the first farmers in Anatolia' *Current Biology* 26.19: 2659–66

Kılınç, G.M., Koptekin, D., Atakuman, Ç., Sümer, A.P., Dönertaş, H.M., Yaka, R., Bilgin, C.C, Büyükkarakaya, A.M., Baird, D., Altınışık, E., Flegontov, P., Götherström, A., Togan, İ., Somel, M. 2017: 'Archaeogenomic analysis of the first steps of Neolithization in Anatolia and the Aegean' *Proceedings of the Royal Society B: Biological Sciences* 284.1867: 2017–64

Knapp, A.B. 2008: *Prehistoric and Protohistoric Cyprus: Identity, Insularity, and Connectivity*. Oxford, Oxford University Press

Kohn, L.A.P., Leigh, S.R., Cheverud, J.M. 1995: 'Asymmetric vault modification in Hopi crania' *American Journal of Physical Anthropology* 98.2: 173–95

Kohn, L.A.P., Leigh, S.R., Jacobs, S.C., Cheverud, J.M. 1993: 'Effects of annular cranial vault modification on the cranial base and face' *American Journal of Physical Anthropology* 90.2: 147–68

Kurth, G., Röhrer-Ertl, O. 1981: 'On the anthropology of the Mesolithic to Chalcolithic human remains from the Tell es-Sultan in Jericho, Jordan' *Excavations at Jericho* 3.1: 409–99

Lambert, P.J. 1979: 'Early Neolithic cranial deformation at Ganj Dareh Tepe, Iran' *Canadian Review of Physical Anthropology* 1.2: 51–54

Lorentz, K.O. 2002: 'Cultures of physical modifications: child bodies in ancient Cyprus' *Bar International Series* 1040: 203–10

— 2008: 'From life course to longue durée: headshaping as gendered capital?' in D. Bolger (ed.) *Gender through Time in the Ancient Near East*. Plymouth, Altamira Press: 281–311

— 2009: 'The malleable body: headshaping in Greece and the surrounding regions' *Hesperia Supplements* 43: 75–98

— 2010: 'Ubaid headshaping: negotiations of identity through physical appearance' in R.A. Carter, G. Philip (eds) *Beyond the Ubaid: Transformation and Integration in the Late Prehistoric Societies of the Middle East 63*. Chicago, Oriental Institute of the University of Chicago: 125–48

— 2017: 'Marking identity through cultural cranial modification within the first sedentary communities (ninth to eighth millennium BCE) in the Near East: Tepe Abdul Hosein, Iran' *International Journal of Osteoarchaeology* 27.6: 973–83

Meiklejohn, C., Agelarakis, A., Akkermans, P.M.M.G., Smith, P.E.L., Solecki, R. 1992: 'Artificial cranial deformation in the Proto-neolithic and Neolithic Near East and its possible origin: evidence from four sites' *Paléorient* 18.2: 83–97

Mellink, M.J., Angel, J.L. 1968: 'Excavations at Karatas-Semayuk in Lycia, 1967' *American Journal of Archaeology* 70.3: 243–63

Miyake, Y. 2010: 'Excavations at Salat Cami Yani 2004–2006: a pottery Neolithic site in the Turkish Tigris Valley' *Proceedings of the 6th International Congress of the Archaeology of the Ancient Near East* 5: 417–29

Molleson, T., Campbell, S. 1995: 'Deformed skulls at Tell Arpachiyah: the social context' *The Archaeology of Death in the Ancient Near East*. Oxford, Oxbow Books: 45–55

Moss, M.L. 1958: 'The pathogenesis of artificial cranial deformation' *American Journal of Physical Anthropology* 16.3: 269–86

Ortner, J. 1996: 'Artificial cranial deformation of a human skull from Choga Mish' *Choga Mish* 1: 319–20

— 2003: *Identification of Pathological Conditions in Human Skeletal Remains*. San Diego, Academic Press

Özbek, M. 1974a: 'A propos des déformations crâniennes artificielles observées au Proche-Orient' *Paléorient* 2.2: 469–76

— 1974b: 'Etude de la déformation cranienne artificielle Chez Les Chalcolithiques de Byblos (Liban). Modifications consécutives sur le crane' *Bulletins et Memoires de La Société d'Anthropologie de Paris* 1.4: 455–81

— 1984: 'Etude anthropologique des restes humains de Hayaz Hoyuk' *Anatolica* 11: 155–70

— 2001: 'Cranial deformation in a subadult sample from Değirmentepe (Chalcolithic, Turkey)' *American Journal of Physical Anthropology: The Official Publication of the American Association of Physical Anthropologists* 115.3: 238–44

Parker, B.J. 2010: 'Networks of interregional interaction during Mesopotamia's Ubaid Period' in R.A. Carter, G. Philip (eds) *Beyond the Ubaid: Transformation and Integration in the Late Prehistoric Societies of the Middle East No. 63*. Chicago, Oriental Institute of the University of Chicago: 339–60

Pomeroy, E., Stock, J.T., Zakrzewski, S.R., Lahr, M.M. 2009: 'A metric study of three types of artificial cranial modification from north-central Peru' *International Journal of Osteoarchaeology* 20.3: 317–34

Saatçioğlu, A. 1982: 'A survey on the racial types of Anatolian skeletal remains' *Ankara Üniversitesi Dil ve Tarih-Coğrafya Fakültesi Dergisi* 30.1–2: 193–209

Şekeryapan, C. 2005: *An Extended Study on the Alu Insertion Polymorphisms in Anatolian Human Population*. PhD thesis, Middle East Technical University, Ankara

Şenyürek, M.S. 1941: 'Anadolu Bakır Çağı ve Eti Sekenesinin kraniyolojik tetkiki' *Belleten* 5.19: 219–53

Şenyürek, M.S., Tunakan, S. 1951: 'Şeyh Höyük iskeletleri' *Belleten* 15: 431–45

Skourtanioti, E., Erdal, Y.S., Frangipane, M., Balossi Restelli, F., Yener, K.A., Pinnock, F., Matthiae, P., Özbal, R., Schoop, U.-D., Guliyev, F., Akhundov, T., Lyonnet, B., Hammer, M.L., Nugent, S.E., Burri, M., Neumann, G.U., Penske, S., Ingman, T., Krause, J. 2020: 'Genomic history of Neolithic to Bronze Age Anatolia, Northern Levant, and Southern Caucasus' *Cell* 181.5: 1158–75

Sørensen, M.L.S. 1997: 'Reading dress: the construction of social categories and identities in Bronze Age Europe' *Journal of European Archaeology* 5.1: 93–114

Soto-Heim, P. 1986: 'Déformation crânienne dans l'Iran ancien' *Bulletins et Mémoires de La Société d'anthropologie de Paris* 3.2: 105–16

Stein, G. 2011: 'Tell Zeidan 2010' *Oriental Institute Annual Report* 2010–2011: 121–38

Tiesler, V. 2014: *The Bioarchaeology of Artificial Cranial Modifications*. New York, Springer Science & Business Media

Trinkaus, E. 1982: 'Artificial cranial deformation in the Shanidar 1 and 5 Neandertals' *Current Anthropology* 23.2: 198–99

van Duijvenbode, A. 2012: 'A reappraisal of intentional cranial modification among the indigenous inhabitants of Suriname in Prehistoric and Historic times' *Caribbean Connections* 2.1: 1–11

7. A genomic history of Neolithisation in Anatolia: preliminary insights

Ayça Aydoğan, N. Ezgi Altınışık, Hasan Can Gemici, Çiğdem Atakuman, Füsun Özer, Mehmet Somel

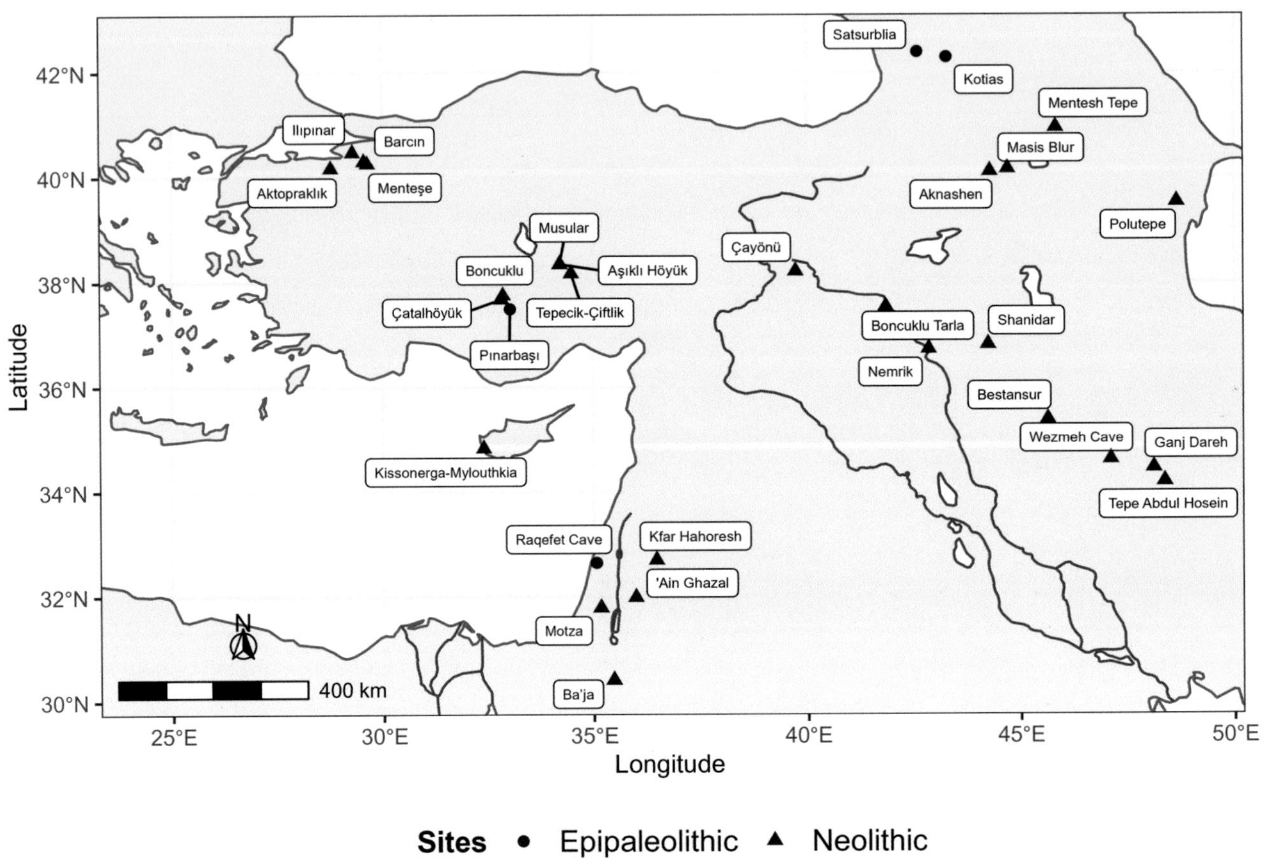

Fig. 7.1. Map showing location of sites with genomic data discussed in the text (map by Ayça Aydoğan).

In this review we provide a summary of recent results from genomic analyses on early Holocene Anatolia. We focus on the Neolithic Transition, namely the shift to a lifestyle based on food production and sedentism that initially emerged in a region encompassing the Levant, Upper Mesopotamia (southeast Anatolia) and the western Zagros mountains in the tenth millennium BC (Asouti, Fuller 2013; Ibáñez et al. 2018). From this 'Fertile Crescent', Neolithic lifeways soon expanded into central Anatolia and Cyprus in the ninth millennium BC, and later started spreading to western Anatolia and southeast Europe in the west, and to the periphery of central Asia and the Caucasus in the east, around the seventh and sixth millennia BC.

Although this overall process has been well described based on material culture evidence over the last 50 years, there have remained unanswered questions regarding both movements of and interactions between Neolithic societies across southwest Asia, and also their social traditions. Some of these are questions that ancient DNA (aDNA) data, in conjunction with archaeological and anthropological evidence, could provide clues for. Specifically, population genetic analyses using aDNA data may allow testing as to whether the spread of Neolithisation occurred mainly via intercultural relationships among communities in southwest Asia, or mainly through human movement. Ancient DNA also provides a means of assessing biological kinship among Neolithic intramural burials, and can thus provide insight into the customs of these societies.

Population structure in southwest Asia before the Neolithic Transition

The late Upper Palaeolithic (UP) and Epipalaeolithic population structure of southwest Asia is represented in aDNA data only by sparse sampling. These include late Upper Palaeolithic and Mesolithic (ca 11,000 and ca 8000 BC, respectively) genomes from the southern Caucasus (Jones et al. 2015) and the Zagros/southern Caspian (Lazaridis et al. 2016; Narasimhan et al. 2019), genomes from the Epipalaeolithic (Natufian) Raqefet Cave (ca 12,000–10,000 BC) from the southern Levant (Lazaridis et al. 2016), and a genome from the Epipalaeolithic phase (ca 15,000 BC) of the Pınarbaşı rock shelter in south-central Anatolia (Feldman et al. 2019). Figure 7.1 shows the locations of the sites from which genomic data have been published to date. Notably, the Pınarbaşı Epipalaeolithic community was not as sedentary as their contemporary Natufians, but their ritual practices, such as skull retrieval, do show resemblances with the latter (Baird et al. 2013).

The analysis of these genomes, in conjunction with genomes from the later Neolithic period sites (detailed below) and from late Upper Palaeolithic and Mesolithic Europe (Lazaridis et al. 2014; Skoglund et al. 2014; Mathieson et al. 2018) has led to the following main observations regarding the region's pre-Neolithic population structure:

- The early Holocene southern Caucasus and Zagros populations are genetically highly similar, suggesting that they share close ancestry and/or remained in close contact through or after the late UP.
- This early Holocene population of the Caucasus and western Zagros, which likely extended across eastern southwest Asia, is genetically distinct from the populations in western southwest Asia, currently represented by populations in the southern Levant and in central Anatolia.
- The Epipalaeolithic genome from Pınarbaşı in central Anatolia is likely a fair representative of the southern and central Anatolian gene pool, although this is still partly speculative. Importantly, the Pınarbaşı individual's genome appears to carry ancestry from Levant and Balkan-related sources (Feldman et al. 2019) and may reflect an admixture event dating to before the Holocene (Marchi et al. 2022).

The latter point is in line with known regional connections, as observed from similarities in ritual-related behaviour between south-central Anatolia and the Levant, and also the early presence of Cappadocian obsidian in the Levant (e.g. Baird et al. 2013; Frahm, Hauck 2017). In turn, the east-west differentiation observed in southwest Asia may partly be due to the impact of the region's mountain ranges on human movement, possibly exacerbated by harsh climatic conditions during the Last Glacial Maximum (LGM).

However, interactions documented during the subsequent Neolithic period across the region (see below) raise doubt on isolation-by-distance being the sole driving force behind east-west differentiation in southwest Asia. Why the Levant-related populations ended up admixing with Balkan/Aegean-related Upper Palaeolithic groups but not their eastern neighbours from the Zagros remains an open question.

The development of Aceramic Neolithic communities

The genetic sampling of Aceramic Neolithic (or Pre-Pottery Neolithic, PPN) period sites from roughly 10,000–7000 BC covers Boncuklu and Aşıklı from south-central Anatolia (Kılınç et al. 2016; Feldman et al. 2019; Yaka et al. 2021), Ain Ghazal, Kfar Hahoresh and Motza from the southern Levant (Lazaridis et al. 2016), and Ganj Dareh and Wezmeh Cave from the central Zagros (Broushaki et al. 2016; Gallego-Llorente et al. 2016; Lazaridis et al. 2016) (fig. 7.1). More recently, genetic data from PPN Upper Mesopotamia, specifically the Tigris area (Çayönü, Boncuklu Tarla, Nemrik) (Altınışık et al. 2022; Lazaridis et al. 2022), as well as the northwest Zagros region (Bestansur and Shanidar) and Cyprus (Kissonerga-Mylouthkia) (Lazaridis et al. 2022), were published. However, some other key regions of early Neolithic southwest Asia, such as the Upper Euphrates region and the northern Levant, remain genetically unsampled. We also lack data on pre-Neolithic populations in west, north and east Anatolia.

The comparative analysis of these Aceramic-period genomes, among themselves and with earlier period samples (see above), has revealed a number of interesting patterns:

- Neolithic development appears to have involved local population continuity in the Levant, the Zagros and central Anatolia. The Levantine PPN groups show the highest genetic affinity to Natufians, and Neolithic Zagros communities to earlier groups that inhabited the same region in the Late Pleistocene and Early Holocene (Lazaridis et al. 2016), while Boncuklu and Aşıklı show the highest genetic affinity to Pınarbaşı (Yaka et al. 2021) relative to any other early

Holocene population. The simplest explanation for this observation is local continuity between the Epipalaeolithic and early Neolithic communities.

- The number of samples from Upper Mesopotamia is much sparser than in central Anatolia, due to worse aDNA preservation. Recently published aDNA data from Upper Mesopotamian PPN populations (e.g. Upper Tigris groups from Çayönü and Boncuklu Tarla) can be genetically modelled as mixtures of Anatolian, Levantine and Zagros-related sources (Altınışık et al. 2022; Lazaridis et al. 2022). The time of these putative admixture events is yet unknown, as is the pre-Neolithic gene pool in this region. The PCA-based comparisons suggest that the proportions of genetic components per population (between Çayönü, Boncuklu Tarla and Nemrik) appear variable, such that the influence of Zagros-related ancestry among the populations in Upper Mesopotamia is greater in the more eastern populations. For instance, Nemrik, geographically closer to the southeast, has a higher genetic association with the Zagros region, but Çayönü has more affinity with Anatolia and the Levant (Altınışık et al. 2022; Lazaridis et al. 2022).
- There is evidence for some degree of interregional human movement between the Epipalaeolithic and the PPN. Feldman and colleagues estimate that there is ca 90% Pınarbaşı ancestry and ca 10% Zagros/Caucasus ancestry in the Boncuklu Höyük population (Feldman et al. 2019). Meanwhile, the southern Levant populations in the PPN show higher affinity to early Holocene Anatolia than the Natufian sample (Lazaridis et al. 2016), and are estimated to carry ca 20% of their ancestry from Anatolia-related sources and ca 80% from Natufians (Feldman et al. 2019). We note that these qpAdm-based admixture proportions should not be taken at face value (i.e. they do not represent the proportion of individuals mixing). Furthermore, it is critical to consider that unknown population structure (i.e. genetic differences within a region) can confound inferences on temporal change. Still, the results are compatible with some degree of interregional movement. Observing similar trends across multiple contemporaneous sites further increases the reliability of these inferences.
- A clearcut case of long-distance movement during the PPN was the settlement of Cyprus by Neolithic groups (Vigne et al. 2012). Although archaeological as well as zooarchaeological and botanical evidence suggest that the Cypriot Neolithic groups were strongly tied to the Levant (Vigne et al. 2012), the presence of central Anatolian obsidian in Cyprus had also hinted at Anatolia-Cyprus connections (Moutsiou 2019). A recent study presenting aDNA from the PPN site of Kissonerga-Mylouthkia in Cyprus found that its strongest genetic association is with central Anatolian groups, although the Cypriot sample also had higher southern Levant ancestry than PPN central Anatolians (Lazaridis et al. 2022). The location of the migrant source population(s) who settled in Cyprus is still, therefore, an open question.
- Analyses focusing on intra-group diversity show that the Boncuklu and Aşıklı genomes from the central Anatolian PPN carry conspicuously low genetic diversity, akin to European and Caucasian Mesolithic foragers, while diversity is higher in Anatolian PN populations (Kılınç et al. 2016; Yaka et al. 2021). This PPN-PN difference in genetic diversity was initially interpreted as a putative signature of foraging-based demography (Kılınç et al. 2016). However, our later analysis showed that data from other PPN sites, such as 'Ain Ghazal and Ganj Dareh, do not display the low diversity signature observed in the central Anatolian PPN (Altınışık et al. 2022). Hence, it is possible that early Holocene population sizes in the southern Levant and the western Zagros, which constitute parts of the original 'Fertile Crescent', were larger relative to population sizes in the Central Anatolian Plateau.

Another interesting point is the low genetic diversity observed in the two PPN communities in central Anatolia. This observation appears consistent with demographic modelling results based on radiocarbon dating data which suggest that populations in south-central Anatolia may have been more adversely affected by the Younger Dryas, and that signs of population increase in the early Holocene appeared late in south-central Anatolia relative to those in the Levant and Mesopotamia (Roberts et al. 2018). We also note that these 'effective population size' estimates, which are inferred from genetic data, can remain high through extended mate exchange networks even in the face of small population sizes per site. Marchi and colleagues (2022) have accordingly speculated that some early sedentary communities that exhibit low diversity, such as those in central Anatolia, might have more limited partner exchange ranges than foragers or other early Neolithic groups; this may again be linked to a smaller regional population size.

The Aceramic-to-Ceramic Neolithic Transition: evidence for human movements into Anatolia

The transition from the Aceramic to the Ceramic (Pottery Neolithic, PN) Neolithic period around 7500–7000 BC, with fully established agriculture and domestic animal husbandry, is currently best observed through ancient genome evidence from central Anatolia. This is due to the fact that samples from multiple sites in the same region have been sequenced from pre-7000 BC settlements (Aşıklı and Boncuklu Höyük) and also post-7000 BC settlements (Çatalhöyük and Tepecik-Çiftlik) (Kılınç et al. 2016; Yaka et al. 2021). In addition, ancient genomes dated to 6500–6000 BC from Barcın Höyük in northwest Anatolia are available (Mathieson et al. 2015; Hofmanová et al. 2016) (Fig. 7.1). Comparative analysis of this data has yielded the following observations:

— Both Çatalhöyük and Tepecik-Çiftlik are highly similar to the other Aceramic communities of central Anatolia, represented by Boncuklu and Aşıklı (Kılınç et al. 2016; Feldman et al. 2019; Yaka et al. 2021). This again suggests population continuity in central Anatolia between the Aceramic and Ceramic Neolithic periods.

— Genomes from Barcın Höyük are also highly similar to those of contemporary central Anatolians such as Çatalhöyük, although the interpretation of this result is less straightforward in the absence of earlier genomes from western Anatolia.

— Early Holocene genomes from the Levant show higher affinity to PN Anatolian genomes than to PPN Anatolian genomes. This speaks for population movement into Anatolia from the south at some time between roughly 7500 and 6500 BC. For instance, Yaka and colleagues estimate ca 10% southern Levant PPN-related ancestry and ca 90% central Anatolian PPN-related ancestry in the Çatalhöyük and Barcın Höyük populations (Yaka et al. 2021). This putative gene flow into central Anatolia from the south could be a factor in why intra-group genetic diversity appears to rapidly increase in the PN Anatolian communities relative to those of the PPN (Yaka et al. 2021).

— In the Tepecik-Çiftlik genome sample some individuals show additional Zagros/Caucasus affinities, although this potential eastern admixture signal is not as conspicuous in the Barcın and Çatalhöyük populations (Kılınç et al. 2016). The Tepecik-Çiftlik sample also stands out among samples from the other studied Neolithic Anatolian communities, as it shows no genetic coherence within the settlement. Specifically, while genomes sampled from Anatolian Neolithic settlements are generally genetically more similar to other genomes from the same settlement, relative to others, this coherence trend is not observed in the Tepecik-Çiftlik sample (Yaka et al. 2021). However, caution is warranted as the sample size is small (n=5).

In summary, the transition to the full-scale agricultural economies of the Ceramic Neolithic in central Anatolia appears to have involved broad-scale population continuity coupled with low-level interregional human movement, as in earlier periods. The connection between the central Anatolian demographic and the Levant is clear, especially considering that it is observed in parallel in two regions of central Anatolia: the Konya Plain and Cappadocia.

A point of interest here is the Zagros/Caucasus influence observed in the Tepecik-Çiftlik sample. Zagros/Caucasus-related ancestry is further observed to increase in genome samples from other sixth- and fifth-millennium sites, such as Büyükkaya and Çamlıbel Tarlası in central Anatolia (Skourtanioti et al. 2020) and Kumtepe in western Anatolia (Kılınç et al. 2016; Omrak et al. 2016). An attractive topic of research is this growing eastern influence in Anatolia between the seventh and fifth millennia, including its tempo (i.e. whether gene flow occurred at specific intervals or gradually), its precise sources, putative sex-biases involved and its possible reflections in material culture. The parallelism between eastern genetic influence in post-Neolithic Anatolia and the west Aegean (i.e. modern-day Greece) (Clemente et al. 2021; Lazaridis et al. 2022) also remains to be investigated.

West Anatolia, the Aegean and the Caucasus: possible human movements out of Anatolia

The spread of Neolithic lifeways westward from central Anatolia into the western range of the peninsula – including the Lakes Region in the southwest, the west Anatolian Aegean coast, and the Marmara region – occurred early in the seventh millennium BC (Özdoğan 2014; Reingruber 2015). This was followed by Neolithic expansion into the Balkans and Europe (Krauss et al. 2017). Traditionally, the process has been described as largely driven by human migration. Indeed, analysing genome-wide variation, European Neolithic farmers form a distinct genetic cluster apart from European Mesolithic hunter-gatherers, indicating a lack of continuity between the Mesolithic and early Neolithic populations in Europe (Skoglund et al. 2014; Lazaridis et al. 2014; 2016; Mathieson et al. 2015; 2018). This is in stark contrast to continuity between Epipalaeolithic and Neolithic groups in central Anatolia and the Fertile Crescent.

In Europe, the incoming population associated with European Neolithic settlements is genetically close to (currently sampled and sequenced) central and northwest Anatolian Neolithic communities (Mathieson et al. 2015; Hofmanová et al. 2016; Kılınç et al. 2016; Chyleński et al. 2019; Marchi et al. 2022), but not to communities from the southern Levant or Balkan Mesolithic groups. This is intriguing, as it argues against a Levantine origin for the sampled Aegean and Balkan Neolithic groups, even though previous research has highlighted the parallels in material culture between western Anatolian Neolithic groups and those from the Levant (Horejs et al. 2015).

The ancestry (or genetic profile) associated with burials from European Neolithic settlements has been termed the 'Anatolian Neolithic' or the 'Anatolian Farmer' (Mathieson et al. 2015). Meanwhile, Kılınç and colleagues (2017) have raised the question of whether the Aegean/Balkan Neolithic communities that drove the Neolithic population movement through the Mediterranean and central Europe had originated in Anatolia. They argued for the possibility that Epipalaeolithic and Mesolithic populations in the Aegean – so far genetically unsampled – might have carried ancestry profiles highly similar to early Holocene central Anatolians. They based their argument on higher genetic diversity in the Aegean Neolithic groups relative to central Anatolians.

In this scenario, local Aegean Mesolithic populations could have adopted Neolithic cultural elements via interactions with their eastern neighbours. If so, the actual population movements that brought Neolithisation to mainland Europe and the European Mediterranean coast may have started from the west Aegean (modern-day Greece) and the Balkans. A direct solution to this problem would be genetic analysis and sequencing of pre-Neolithic populations in western Anatolia and the Aegean. Meanwhile, Marchi et al. (2022) have recently performed detailed modelling of high-quality genomes from the early Holocene northern Balkans and Anatolia. Their models suggest that European Neolithic communities had derived from Aegean/Balkan groups (e.g. Nea Nikomedeia), who in turn had split from central Anatolian Neolithic groups (e.g. Boncuklu) much earlier than 7000 BC and the westward expansion of Neolithic lifeways in Anatolia. This appears inconsistent with a simple model of central Anatolian Neolithic populations colonising the Aegean. Still, some degree of population movement between central and western Anatolia post-7000 BC is not unlikely.

In parallel with the westwards spread of Neolithic lifeways, the late Neolithic and Chalcolithic periods in Anatolia also appear to have involved population interactions with the east. This not only included eastern gene flow into central Anatolia, but also central Anatolian-related ancestry appearing in the Caucasus and Zagros regions in genome samples post-6000 BC (Lazaridis et al. 2016; Wang et al. 2019; Lazaridis et al. 2022). We note that sparse temporal and spatial sampling in the east leaves open the possibility of population movements in earlier millennia that could have involved groups genetically related to those of central Anatolia spreading eastwards.

Social structure of Neolithic communities

Other than demographic inference, aDNA can be employed to study questions about social traditions in past societies. The insights such studies can provide into prehistoric social structures may be particularly valuable, but such work is currently rare (reviewed in Vai et al. 2020). Recently, Yaka et al. (2021) studied genetic kinship among individuals co-buried within the same or neighbouring buildings in five Neolithic Anatolian settlements: Aşıklı and Boncuklu Höyük of Aceramic Neolithic central Anatolia, Çatalhöyük and Tepecik-Çiftlik of Ceramic Neolithic central Anatolia and Barcın Höyük of Ceramic Neolithic northwest Anatolia. Of the 50 intramural co-buried pairs they analysed, they identified a total of eight first- and second-degree related pairs. Interestingly, among co-burials in Aceramic Neolithic buildings, which represent the first permanent settlements on the Anatolian Plateau, the proportion of genetic relatives among co-burials was relatively high. In contrast, co-burials from later Çatalhöyük and Barcın buildings included fewer close genetic relatives.

More recently, Altınışık et al. (2022) presented kinship estimates among nine co-burials across three PPN buildings from Çayönü in the Upper Tigris region. Again, the majority of the studied individuals (7/9) had at least one close genetic kin, in line with contemporaneous Aşıklı and Boncuklu Höyük from central Anatolia.

The small sample sizes do not allow for a definitive conclusion regarding possible temporal shifts in co-burial patterns and social structure between Aceramic/PPN and Ceramic Neolithic period communities. It is nevertheless intriguing that earlier work on Çatalhöyük co-burials that were based on dental metric and non-metric traits (Pilloud, Larsen 2011) and on mitochondrial DNA (Chyleński et al. 2019) also suggested a lack of close biological kinship. This result has been interpreted as an indication that in Neolithic Çatalhöyük and elsewhere, social organisation was not exclusively based on biological family ties (Yaka et al. 2021).

Future directions

The picture provided here already includes some notable novel patterns that had remained contentious until less than a decade ago, such as heterogeneous dynamics with respect to the mode of the Neolithic dissemination in west Eurasia. The movement of domestic crops, animals or novel technologies *within* southwest Asia does not appear to have involved colonisation-style mass population movements. In contrast, the Neolithisation of Europe does appear to have involved mass movement, with likely origins around the Aegean. The process in the Caucasus is less clear, but some degree of population movement from Anatolia-related sources and rapid admixture with local populations appears to be a likely scenario.

We anticipate that the picture will be further clarified as more ancient genomes are produced, both from Neolithic and also Epipalaeolithic sites. For instance, genomes from southeast Anatolia could reveal the association among communities in the Fertile Crescent and central Anatolia, and also help to better understand associations among the societies in the Fertile Crescent. Genomes from the pre-7000 BC Aegean could help resolve the question of how the Aegean Neolithisation process started. Although these are tempting questions, we note that the warm climatic conditions of both southeast Anatolia and the Aegean mean obtaining aDNA from these regions will pose more significant challenges than in central Anatolia and in most parts of Europe.

Eventually, though, sampling will become denser, which will allow for more comprehensive and detailed mapping of genetic-level population interactions in space and time. Comparing these interactions with cultural dynamics as captured by material culture patterns, as well as palaeoclimate data, poses the next challenge. Developing frameworks for joint analysis of quantitative material culture or climate data with population genetic data to test specific hypotheses about cultural changes and human movements may be interesting paths to follow.

Acknowledgements
This work was supported by a European Research Council Consolidator Grant H2020 ERC (no. 772390 'EOGENE' to Mehmet Somel).

Bibliography

Altınışık, N.E., Kazancı, D.D., Aydoğan, A., Gemici, H.C., Erdal, Ö.D., Sarıaltun, S., Vural, K.B., Koptekin, D., Gürün, K., Sağlıcan, E., Çakan, G., Koruyucu, M.M., Kempe Lagerholm, V., Karamurat, C., Özkan, M., Kılınç, G.M., Sevdar, A., Sürer, E., Götherström, A., Atakuman, Ç., Erdal, Y.S., Özer, F., Erim-Özdoğan, A., Somel, M. 2022: 'A genomic snapshot of demographic and cultural dynamism in Upper Mesopotamia during the Neolithic transition' *bioRxiv*. https://www.biorxiv.org/content/10.1101/2022.01.31.478487v1

Arbuckle, B.S. 2014: 'Pace and process in the emergence of animal husbandry in Neolithic Southwest Asia' *Bioarchaeology of the Near East* 8: 1–29

Asouti, E., Fuller, D.Q. 2013: 'A contextual approach to the emergence of agriculture in Southwest Asia: Reconstructing Early Neolithic plant-food production' *Current Anthropology* 54.3: 299–345. https://doi.org/10.1086/670679

Baird, D., Asouti, E., Astruc, L., Baysal, A., Baysal, E., Carruthers, D., Fairbairn, A., Kabukcu, C., Jenkins, E., Lorentz, K., Middleton, C., Pearson, J., Pirie, A. 2013: 'Juniper smoke, skulls and wolves' tails. The Epipalaeolithic of the Anatolian Plateau in its South-West Asian context; insights from Pınarbaşı' *Levant* 45.2: 175–209. https://doi.org/10.1179/0075891413Z.00000000024

Baird, D., Fairbairn, A., Jenkins, E., Martin, L., Middleton, C., Pearson, J., Asouti, E., Edwards, Y., Kabukcu, C., Mustafaoglu, G., Russell, N., Bar-Yosef, O., Jacobsen, G., Wu, X., Baker, A., Elliott, S. 2018: 'Agricultural origins on the Anatolian Plateau' *Proceedings of the National Academy of Sciences of the United States of America* 115.14: E3077–86. https://doi.org/10.1073/pnas.1800163115

Broushaki, F., Thomas, M.G., Link, V., López, S., van Dorp, L., Kirsanow, K., Hofmanová, Z., Diekmann, Y., Cassidy, L.M., Díez-del-Molino, D., Kousathanas, A., Sell, C., Robson, H.K., Martiniano, R., Blöcher, J., Scheu, A., Kreutzer, S., Bollongino, R., Bobo, D., Davoudi, H., Munoz, O., Currat, M., Shennan, S., Veeramah, K.R., Mashkour, M., Wegmann, D. 2016: 'Early Neolithic genomes from the Eastern Fertile Crescent' *Science* 353.6298: 499–504. https://doi.org/10.6084/m9.figshare.3470135

Chyleński, M., Ehler, E., Somel, M., Yaka, R., Krzewińska, M., Dabert, M., Juras, A., Marciniak, A. 2019: 'Ancient mitochondrial genomes reveal the absence of maternal kinship in the burials of Çatalhöyük people and their genetic affinities' *Genes* 10.3: 1–14. https://doi.org/10.3390/genes10030207

Daly, K.G., Delser, P.M., Mullin, V.E., Scheu, A., Mattiangeli, V., Teasdale, M.D., Hare, A.J., Burger, J., Verdugo, M.P., Collins, M.J., Kehati, R., Erek, C.M. 2018: 'Ancient goat genomes reveal mosaic domestication in the Fertile Crescent' *Science* 361: 85–88

Düring, B.S. (ed.) 2011: *The Prehistory of Asia Minor: From Complex Hunter-Gatherers to Early Urban Societies*. Cambridge, Cambridge University Press

Feldman, M., Fernández-Domínguez, E., Reynolds, L., Baird, D., Pearson, J., Hershkovitz, I., May, H., Goring-Morris, N., Benz, M., Gresky, J., Bianco, R.A., Fairbairn, A., Mustafaoğlu, G., Stockhammer, P.W., Posth, C., Haak, W., Jeong, C., Krause, J. 2019: 'Late Pleistocene human genome suggests a local origin for the first farmers of Central Anatolia' *Nature Communications* 10.1: 1–10. https://doi.org/10.1038/s41467-019-09209-7

Frahm, E., Hauck, T.C. 2017: 'Origin of an obsidian scraper at Yabroud Rockshelter II (Syria): implications for Near Eastern social networks in the Early Upper Palaeolithic' *Journal of Archaeological Science: Reports* 13: 415–27. https://doi.org/10.1016/j.jasrep.2017.04.021

Fuller, D.Q., Willcox, G., Allaby, R.G. 2011: 'Cultivation and domestication had multiple origins: arguments against the Core Area hypothesis for the origins of agriculture in the Near East' *World Archaeology* 43.4: 628–52. https://doi.org/10.1080/00438243.2011.624747

Gallego-Llorente, M., Connell, S., Jones, E.R., Merrett, D.C., Jeon, Y., Eriksson, A., Siska, V., Gamba, C., Meiklejohn, C., Beyer, R., Jeon, S., Cho, Y.S., Hofreiter, M., Bhak, J., Manica, A., Pinhasi, R. 2016: 'The genetics of an Early Neolithic pastoralist from the Zagros, Iran' *Scientific Reports* 6. https://doi.org/10.1038/srep31326

Hofmanová, Z., Kreutzer, S., Hellenthal, G., Sell, C., Diekmann, Y., Díez-del-Molino, D., van Dorp, L. López, S., Kousathanas, A., Link, V., Kirsanow, K., Cassidy, L.M., Martiniano, R., Strobel, M., Scheu, A., Kotsakis, K., Halstead, P., Triantaphyllou, S., Kyparissi-Apostolika, N., Urem-Kotsou, D., Ziota, C., Adaktylou, F., Gopalan, S., Bobo, D.M., Winkelbach, L., Blöcher, J., Unterländer, M., Leuenberger, C., Çilingiroğlu, Ç., Horejs, B., Gerritsen, F., Shennan, S.J., Bradley, D.G., Currat, M., Veeramah, K.R., Wegmann, D., Thomas, M.G., Papageorgopoulou, C., Burger, J. 2016: 'Early farmers from across Europe directly descended from Neolithic Aegeans' *Proceedings of the National Academy of Sciences of the United States of America* 113.25: 6886–91. https://doi.org/10.1073/pnas.1523951113

Horejs, B., Milić, B., Ostmann, F., Thanheiser, U., Weninger, B., Galik, A. 2015: 'The Aegean in the early 7th Millennium BC: Maritime networks and colonization' *Journal of World Prehistory* 28.4: 289–330. https://doi.org/10.1007/s10963-015-9090-8

Ibáñez, J.J., González-Urquijo, J., Teira-Mayolini, L.C., Lazuén, T. 2018: 'The emergence of the Neolithic in the Near East: a protracted and multi-regional mode' *Quaternary International* 470: 226–52. https://doi.org/10.1016/j.quaint.2017.09.040

Ibáñez, J.J., Ortega, D., Campos, D., Khalidi, L., Méndez, V., Teira, L. 2016: 'Developing a complex network model of obsidian exchange in the Neolithic Near East: linear regressions, ethnographic models and archaeological data' *Paléorient* 42.2: 9–32. https://doi.org/10.3406/paleo.2016.5718

Jones, E.R., Gonzalez-Fortes, G., Connell, S., Siska, V., Eriksson, A., Martiniano, R., McLaughlin, R.L., Gallego Llorente, M., Cassidy, L.M., Gamba, C., Meshveliani, T., Bar-Yosef, O., Müller, W., Belfer-Cohen, A., Matskevich, Z., Jakeli, N., Higham, T.F.G., Currat, M., Lordkipanidze, D., Hofreiter, M., Manica, A., Pinhasi, R., Bradley, D.G. 2015: 'Upper Palaeolithic genomes reveal deep roots of modern Eurasians' *Nature Communications* 6: 1–8. https://doi.org/10.1038/ncomms9912

Kılınç, G.M., Koptekin, D., Atakuman, Ç., Sümer, A.P., Dönertas, H.M., Yaka, R., Bilgin, C.C., Büyükkarakaya, A.M., Baird, D., Altınışık, E., Flegontov, P., Götherström, A., Togan, İ., Somel, M. 2017: 'Archaeogenomic analysis of the first steps of Neolithisation in Anatolia and the Aegean' *Proceedings of the Royal Society B: Biological Sciences* 284.1867. https://doi.org/10.1098/rspb.2017.2064

Kılınç, G.M., Omrak, A., Özer, F., Günther, T., Büyükkarakaya, A.M., Bıçakçı, E., Baird, D., Dönertaş, H.M., Ghalichi, A., Yaka, R., Koptekin, D., Açan, S.C., Parvizi, P., Krzewińska, M., Daskalaki, E.A., Yüncü, E., Dağtaş, N.D., Fairbairn, A., Pearson, J., Mustafaoğlu, G., Erdal, Y.S., Çakan, Y.G., Togan, İ., Somel, M., Storå, J., Jakobsson, M., Götherström, A. 2016: 'The demographic development of the first farmers in Anatolia' *Current Biology* 26.19: 2659–66. https://doi.org/10.1016/j.cub.2016.07.057

Krauss, R., Marinova, E., Brue, H., Weninger, B. 2017: 'The rapid spread of early farming from the Aegean into the Balkans via the Sub-Mediterranean-Aegean Vegetation Zone' *Quaternary International* 496 (March). https://doi.org/10.1016/j.quaint.2017.01.019

Lazaridis, I., Alpaslan-Roodenberg, S., Acar, A., Açıkkol, A., Agelarakis, A., Aghikyan, L., Akyüz, U., Andreeva, D., Andrijašević, G., Antonović, D., Armit, I., Atmaca, A., Avetisyan, P., Aytek, A.İ., Bacvarov, K., Badalyan, R., Bakardzhiev, S., Balen, J., Bejko, L., Bernardos, R., Bertsatos, A., Biber, H., Bilir, A., Bodružić, M., Bonogofsky, M., Bonsall, C., Borić, D., Borovinić, N., Bravo Morante, G., Buttinger, K., Callan, K., Candilio, F., Carić, M., Cheronet, O., Chohadzhiev, S., Chovalopoulou, M.-E., Chryssoulaki, S., Ciobanu, I., Čondić, N., Constantinescu,

M., Cristiani, E., Culleton, B.J., Curtis, E., Davis, J., Demcenco, T.I., Dergachev, V., Derin, Z., Deskaj, S., Devejyan, S., Djordjević, V., Duffett Carlson, K.S., Eccles, L.R., Elenski, N., Engin, A., Erdoğan, N., Erir-Pazarcı, S., Fernandes, D.M., Ferry, M., Freilich, S., Frînculeasa, A., Galaty, M.L., Gamarra, B., Gasparyan, B., Gaydarska, B., Genç, E., Gültekin, T., Gündüz, S., Hajdu, T., Heyd, V., Hobosyan, S., Hovhannisyan, N., Iliev, I., Iliev, L., Iliev, S., İvgin, İ., Janković, I., Jovanova, L., Karkanas, P., Kavaz-Kındığılı, B., Kaya, E.H., Keating, D., Kennett, D.J., Deniz Kesici, S., Khudaverdyan, A., Kiss, K., Kılıç, S., Klostermann, P., Kostak Boca Negra Valdes, S., Kovačević, S., Krenz-Niedbała, M., Krznarić Škrivanko, M., Kurti, R., Kuzman, P., Lawson, A.M., Lazar, C., Leshtakov, K., Levy, T.E., Liritzis, I., Lorentz, K.O., Łukasik, S., Mah, M., Mallick, S., Mandl, K., Martirosyan-Olshansky, K., Matthews, R., Matthews, W., McSweeney, K., Melikyan, V., Micco, A., Michel, M., Milašinović, L., Mittnik, A., Monge, J.M., Nekhrizov, G., Nicholls, R., Nikitin, A.G., Nikolov, V., Novak, M., Olalde, I., Oppenheimer, J., Osterholtz, A., Özdemir, C., Özdoğan, K.T., Öztürk, N., Papadimitriou, N., Papakonstantinou, N., Papathanasiou, A., Paraman, L., Paskary, E.G., Patterson, N., Petrakiev, I., Petrosyan, L., Petrova, V., Philippa-Touchais, A., Piliposyan, A., Pocuca Kuzman, N., Potrebica, H., Preda-Bălănică, B., Premužić, Z., Price, T.D., Qiu, L., Radović, S., Raeuf Aziz, K., Rajić Šikanjić, P., Rasheed Raheem, K., Razumov, S., Richardson, A., Roodenberg, J., Ruka, R., Russeva, V., Şahin, M., Şarbak, A., Savaş, E., Schattke, C., Schepartz, L., Selçuk, T., Sevim-Erol, A., Shamoon-Pour, M., Shephard, H.M., Sideris, A., Simalcsik, A., Simonyan, H., Sinika, V., Sirak, K., Sirbu, G., Šlaus, M., Soficaru, A., Söğüt, B., Sołtysiak, A., Sönmez-Sözer, Ç., Stathi, M., Steskal, M., Stewardson, K., Stocker, S., Suata-Alpaslan, F., Suvorov, A., Szécsényi-Nagy, A., Szeniczey, T., Telnov, N., Temov, S., Todorova, N., Tota, U., Touchais, G., Triantaphyllou, S., Türker, A., Ugarković, M., Valchev, T., Veljanovska, F., Videvski, Z., Virag, C., Wagner, A., Walsh, S., Włodarczak, P., Workman, J.N., Yardumian, A., Yarovoy, E., Yavuz, A.Y., Yılmaz, H., Zalzala, F., Zettl, A., Zhang, Z., Çavuşoğlu, R., Rohland, N., Pinhasi, R., Reich, D. 2022: 'Ancient DNA from Mesopotamia suggests distinct Pre-Pottery and Pottery Neolithic migrations into Anatolia' *Science* 377.6609: 982–87. https://doi.org/10.1126/science.abq0762

Lazaridis, I., Nadel, D., Rollefson, G., Merrett, D.C., Rohland, N., Mallick, S., Fernandes, D., Novak, M., Gamarra, B., Sirak, K., Connell, S., Stewardson, K., Harney, E., Fu, Q., Gonzalez-Fortes, G., Jones, E.R., Roodenberg, S.A., Lengyel, G., Bocquentin, F., Gasparian, B., Monge, J.M., Gregg, M., Eshed, V., Mizrahi, A.S., Meiklejohn, C., Gerritsen, F., Bejenaru, L., Blüher, M., Campbell, A., Cavalleri, G., Comas, D., Froguel, P., Gilbert, E., Kerr, S.M., Kovacs, P., Krause, J., McGettigan, D., Merrigan, M., Merriwether, D.A., O'Reilly, S., Richards, M.B., Semino, O., Shamoon-Pour, M., Stefanescu, G., Stumvoll, M., Tönjes, A., Torroni, A., Wilson, J.F., Yengo, L., Hovhannisyan, N.A., Patterson, N., Pinhasi, R., Reich, D. 2016: 'Genomic insights into the origin of farming in the Ancient Near East' *Nature* 536.7617: 419–24. https://doi.org/10.1038/nature19310

Lazaridis, I., Patterson, N., Mittnik, A., Renaud, G., Mallick, S., Kirsanow, K., Sudmant, P.H., Schraiber, J.G., Castellano, S., Lipson, M., Berger, B., Economou, C., Bollongino, R., Fu, Q., Bos, K.I., Nordenfelt, S., Li, H., Filippo, C. De, Prüfer, K., Sawyer, S., Posth, C., Haak, W., Hallgren, F., Fornander, E., Rohland, N., Delsate, D., Francken, M., Guinet, J.M., Wahl, J., Ayodo, G., Babiker, H.A., Bailliet, G., Balanovska, E., Balanovsky, O., Barrantes, R., Bedoya, G., Ben-Ami, H., Bene, J., Berrada, F., Bravi, C.M., Brisighelli, F., Busby, G.B.J., Cali, F., Churnosov, M., Cole, D.E.C., Corach, D., Damba, L., Driem, G. Van, Dryomov, S., Dugoujon, J.M., Fedorova, S.A., Gallego Romero, I., Gubina, M., Hammer, M., Henn, B.M., Hervig, T., Hodoglugil, U., Jha, A.R., Karachanak-Yankova, S., Khusainova, R., Khusnutdinova, E., Kittles, R., Kivisild, T., Klitz, W., Kučinskas, V., Kushniarevich, A., Laredj, L., Litvinov, S., Loukidis, T., Mahley, R.W., Melegh, B., Metspalu, E., Molina, J., Mountain, J., Näkkäläjärvi, K., Nesheva, D., Nyambo, T., Osipova, L., Parik, J., Platonov, F., Posukh, O., Romano, V., Rothhammer, F., Rudan, I., Ruizbakiev, R., Sahakyan, H., Sajantila, A., Salas, A., Starikovskaya, E.B., Tarekegn, A., Toncheva, D., Turdikulova, S., Uktveryte, I., Utevska, O., Vasquez, R., Villena, M., Voevoda, M., Winkler, C.A., Yepiskoposyan, L., Zalloua, P., Zemunik, T., Cooper, A., Capelli, C., Thomas, M.G., Ruiz-Linares, A., Tishkoff, S.A., Singh, L., Thangaraj, K., Villems, R., Comas, D., Sukernik, R., Metspalu, M., Meyer, M., Eichler, E.E., Burger, J., Slatkin, M., Pääbo, S., Kelso, J., Reich, D., Krause, J. 2014: 'Ancient human genomes suggest three ancestral populations for present-day Europeans' *Nature* 513.7518: 409–13. https://doi.org/10.1038/nature13673

Maeda, O., Lucas, L., Silva, F., Tanno, K.-I., Fuller, D.Q. 2016: 'Narrowing the harvest: increasing sickle investment and the rise of domesticated cereal agriculture in the Fertile Crescent' *Quaternary Science Reviews* 145: 226–37. https://doi.org/10.1016/j.quascirev.2016.05.032

Marchi, N., Winkelbach, L., Schulz, I., Brami, M., Hofmanová, Z., Blöcher, J., Reyna-Blanco, C.S., Diekmann, Y., Thiéry, A., Kapopoulou, A., Link, V., Piuz, V., Kreutzer, S., Figarska, S.M., Ganiatsou, E., Pukaj, A., Struck, T.J., Gutenkunst, R.N., Karul, N., Gerritsen, F., Pechtl, J., Peters, J., Zeeb-Lanz, A., Lenneis, E., Teschler-Nicola, M.,

Triantaphyllou, S., Stefanović, S., Papageorgopoulou, C., Wegmann, D., Burger, J., Excoffier, L. 2022: 'The genomic origins of the world's first farmers' *Cell*. https://doi.org/10.1016/j.cell.2022.04.008

Mathieson, I., Alpaslan-Roodenberg, S., Posth, C., Szécsényi-Nagy, A., Rohland, N., Mallick, S., Olalde, I., Broomandkhoshbacht, N., Candilio, F., Cheronet, O., Fernandes, D., Ferry, M., Gamarra, B., Fortes, G.G., Haak, W., Harney, E., Jones, E., Keating, D., Krause-Kyora, B., Kucukkalipci, I., Michel, M., Mittnik, A., Nägele, K., Novak, M., Oppenheimer, J., Patterson, N., Pfrengle, S., Sirak, K., Stewardson, K., Vai, S., Alexandrov, S., Alt, K.W., Andreescu, R., Antonović, D., Ash, A., Atanassova, N., Bacvarov, K., Gusztáv, M.B., Bocherens, H., Bolus, M., Boroneanţ, A., Boyadzhiev, Y., Budnik, A., Burmaz, J., Chohadzhiev, S., Conard, N.J., Cottiaux, R., Čuka, M., Cupillard, C., Drucker, D.G., Elenski, N., Francken, M., Galabova, B., Ganetsovski, G., Gély, B., Hajdu, T., Handzhyiska, V., Harvati, K., Higham, T., Iliev, S., Janković, I., Karavanić, I., Kennett, D.J., Komšo, D., Kozak, A., Labuda, D., Lari, M., Lazar, C., Leppek, M., Leshtakov, K., Lo Vetro, D., Los, D., Lozanov, I., Malina, M., Martini, F., McSweeney, K., Meller, H., Mentušić, M., Mirea, P., Moiseyev, V., Petrova, V., Douglas Price, T., Simalcsik, A., Sineo, L., Šlaus, M., Slavchev, V., Stanev, P., Starović, A., Szeniczey, T., Talamo, S., Teschler-Nicola, M., Thevenet, C., Valchev, I., Valentin, F., Vasilyev, S., Veljanovska, F., Venelinova, S., Veselovskaya, E., Viola, B., Virag, C., Zaninović, J., Zaüner, S., Stockhammer, P.W., Catalano, G., Krauß, R., Caramelli, D., Zariņą, G., Gaydarska, B., Lillie, M., Nikitin, A.G., Potekhina, I., Papathanasiou, A., Borić, D., Bonsall, C., Krause, J., Pinhasi, R., Reich, D. 2018: 'The genomic history of southeastern Europe' *Nature* 555.7695: 197–203. https://doi.org/10.1038/nature25778

Mathieson, I., Lazaridis, I., Rohland, N., Mallick, S., Patterson, N., Roodenberg, S.A., Harney, E., Stewardson, K., Fernandes, D., Novak, M., Sirak, K., Gamba, C., Jones, E.R., Llamas, B., Dryomov, S., Pickrell, J., Arsuaga, J.L., de Castro, J.M.B., Carbonell, E., Gerritsen, F., Khokhlov, A., Kuznetsov, P., Lozano, M., Meller, H., Mochalov, O., Moiseyev, V., Guerra, M.A.R., Roodenberg, J., Vergès, J.M., Krause, J., Cooper, A., Alt, K.W., Brown, D., Anthony, D., Lalueza-Fox, C., Haak, W., Pinhasi, R., Reich, D. 2015: 'Genome-wide patterns of selection in 230 Ancient Eurasians' *Nature* 528.7583: 499–503. https://doi.org/10.1038/nature16152

Moutsiou, T. 2019: 'A compositional study (pXRF) of Early Holocene Obsidian assemblages from Cyprus, Eastern Mediterranean' *Open Archaeology* 5.1: 155–66. https://doi.org/10.1515/opar-2019-0011

Narasimhan, V.M., Patterson, N., Moorjani, P., Rohland, N., Bernardos, R., Mallick, S., Lazaridis, I., Nakatsuka, N., Olalde, I., Lipson, M., Kim, A.M., Olivieri, L.M., Coppa, A., Vidale, M., Mallory, J., Moiseyev, V., Kitov, E., Monge, J., Adamski, N., Alex, N., Broomandkhoshbacht, N., Candilio, F., Callan, K., Cheronet, O., Culleton, B.J., Ferry, M., Fernandes, D., Freilich, S., Gamarra, B., Gaudio, D., Hajdinjak, M., Harney, É., Harper, T.K., Keating, D., Lawson, A.M., Mah, M., Mandl, K., Michel, M., Novak, M., Oppenheimer, J., Rai, N., Sirak, K., Slon, V., Stewardson, K., Zalzala, F., Zhang, Z., Akhatov, G., Bagashev, A.N., Bagnera, A., Baitanayev, B., Bendezu-Sarmiento, J., Bissembaev, A.A., Bonora, G.L., Chargynov, T.T., Chikisheva, T., Dashkovskiy, P.K., Derevianko, A., Dobeš, M., Douka, K., Dubova, N., Duisengali, M.N., Enshin, D., Epimakhov, A., Fribus, A.V., Fuller, D., Goryachev, A., Gromov, A., Grushin, S.P., Hanks, B., Judd, M., Kazizov, E., Khokhlov, A., Krygin, A.P., Kupriyanova, E., Kuznetsov, P., Luiselli, D., Maksudov, F., Mamedov, A.M., Mamirov, T.B., Meiklejohn, C., Merrett, D.C., Micheli, R., Mochalov, O., Mustafokulov, S., Nayak, A., Pettener, D., Potts, R., Razhev, D., Rykun, M., Sarno, S., Savenkova, T.M., Sikhymbaeva, K., Slepchenko, S.M., Soltobaev, O.A., Stepanova, N., Svyatko, S., Tabaldiev, K., Teschler-Nicola, M., Tishkin, A.A., Tkachev, V.V., Vasilyev, S., Velemínský, P., Voyakin, D., Yermolayeva, A., Zahir, M., Zubkov, V.S., Zubova, A., Shinde, V.S., Lalueza-Fox, C., Meyer, M., Anthony, D., Boivin, N., Thangaraj, K., Kennett, D.J., Frachetti, M., Pinhasi, R., Reich, D. 2019: 'The formation of human populations in South and Central Asia' *Science* 365.6457. https://doi.org/10.1126/science.aat7487

Omrak, A., Günther, T., Valdiosera, C., Svensson, E.M., Malmström, H., Kiesewetter, H., Aylward, W., Storå, J., Jakobsson, M., Götherström, A. 2016: 'Genomic evidence establishes Anatolia as the source of the European Neolithic gene pool' *Current Biology* 26.2: 270–75. https://doi.org/10.1016/j.cub.2015.12.019

Özdoğan, M. 2014: 'A new look at the introduction of the Neolithic way of life in Southeastern Europe. Changing paradigms of the expansion of the Neolithic way of life' Documenta Praehistorica 41.0: 33–49. https://doi.org/10.4312/dp.41.2

Pilloud, M.A., Larsen, C.S. 2011: '"Official" and "Practical" kin: inferring social and community structure from dental phenotype at Neolithic Çatalhöyük, Turkey' *American Journal of Physical Anthropology* 145.4: 519–30. https://doi.org/10.1002/ajpa.21520

Reingruber, A. 2015: 'Preceramic, Aceramic or Early Ceramic? The radiocarbon dated beginning of the Neolithic in the Aegean' *Documenta Praehistorica* 42 (December): 147. https://doi.org/10.4312/dp.42.9

Riehl, S., Zeidi, M., Conard, N.J. 2013: 'Emergence of agriculture in the foothills of the Zagros Mountains of Iran' *Science* 341.6141: 65 LP – 67. https://doi.org/10.1126/science.1236743

Roberts, N., Woodbridge, J., Bevan, A., Palmisano, A., Shennan, S., Asouti, E. 2018: 'Human responses and non-responses to climatic variations during the Last Glacial-Interglacial transition in the Eastern Mediterranean' *Quaternary Science Reviews* 184: 47–67. https://doi.org/10.1016/j.quascirev.2017.09.011

Skoglund, P., Sjödin, P., Skoglund, T., Lascoux, M., Jakobsson, M. 2014: 'Investigating population history using temporal genetic differentiation' *Molecular Biology and Evolution* 31.9: 2516–27. https://doi.org/10.1093/molbev/msu192

Skourtanioti, E., Erdal, Y.S., Frangipane, M., Balossi Restelli, F., Yener, K.A., Pinnock, F., Matthiae, P., Özbal, R., Schoop, U.-D., Guliyev, F., Akhundov, T., Lyonnet, B., Hammer, E.L., Nugent, S.E., Burri, M., Neumann, G.U., Penske, S., Ingman, T., Akar, M., Shafiq, R., Palumbi, G., Eisenmann, S., D'Andrea, M., Rohrlach, A.B., Warinner, C., Jeong, C., Stockhammer, P.W., Haak, W., Krause, J. 2020: 'Genomic history of Neolithic to Bronze Age Anatolia, Northern Levant, and Southern Caucasus' *Cell* 181.5: 1158–1175.e28. https://doi.org/10.1016/j.cell.2020.04.044

Vai, S., Amorim, C.E.G., Lari, M., Caramelli, D. 2020: 'Kinship determination in archeological contexts through DNA analysis' *Frontiers in Ecology and Evolution* 8 (March): 1–9. https://doi.org/10.3389/fevo.2020.00083

Vigne, J.D., Briois, F., Zazzo, A., Willcox, G., Cucchi, T., Thiébault, S., Carrère, I., Franel, Y., Touquet, R., Martin, C., Moreau, C., Comby, C., Guilaine, J. 2012: 'First wave of cultivators spread to Cyprus at least 10,600 y ago' *Proceedings of the National Academy of Sciences of the United States of America* 109.22: 8445–49. https://doi.org/10.1073/pnas.1201693109

Wang, C.-C., Reinhold, S., Kalmykov, A., Wissgott, A., Brandt, G., Jeong, C., Cheronet, O., Ferry, M., Harney, E., Keating, D., Mallick, S., Rohland, N., Stewardson, K., Kantorovich, A.R., Maslov, V.E., Petrenko, V.G., Erlikh, V.R., Atabiev, B.C., Magomedov, R.G., Kohl, P.L., Alt, K.W., Pichler, S.L., Gerling, C., Meller, H., Vardanyan, B., Yeganyan, L., Rezepkin, A.D., Mariaschk, D., Berezina, N., Gresky, J., Fuchs, K., Knipper, C., Schiffels, S., Balanovska, E., Balanovsky, O., Mathieson, I., Higham, T., Berezin, Y.B., Buzhilova, A., Trifonov, V., Pinhasi, R., Belinskij, A.B., Reich, D., Hansen, S., Krause, J., Haak, W. 2019: 'Ancient human genome-wide data from a 3000-year interval in the Caucasus corresponds with eco-geographic regions' *Nature Communications* 10.1. https://doi.org/10.1038/s41467-018-08220-8

Watkins, T. 2008: 'Supra-regional networks in the Neolithic of Southwest Asia' *Journal of World Prehistory* 21.2: 139–71. https://doi.org/10.1007/s10963-008-9013-z

Yaka, R., Mapelli, I., Kaptan, D., Doğu, A., Chyleński, M., Erdal, Ö.D., Koptekin, D., Vural, K.B., Bayliss, A., Mazzucato, C., Fer, E., Çokoğlu, S.S., Lagerholm, V.K., Krzewińska, M., Karamurat, C., Gemici, H.C., Sevkar, A., Dağtaş, N.D., Kılınç, G.M., Adams, D., Munters, A.R., Sağlıcan, E., Milella, M., Schotsmans, E.M.J., Yurtman, E., Çetin, M., Yorulmaz, S., Altınışık, N.E., Ghalichi, A., Juras, A., Bilgin, C.C., Günther, T., Storå, J., Jakobsson, M., de Kleijn, M., Mustafaoğlu, G., Fairbairn, A., Pearson, J., Togan, İ., Kayacan, N., Marciniak, A., Larsen, C.S., Hodder, I., Atakuman, Ç., Pilloud, M., Sürer, E., Gerritsen, F., Özbal, R., Baird, D., Erdal, Y.S., Duru, G., Özbaşaran, M., Haddow, S.D., Knüsel, C.J., Götherström, A., Özer, F., Somel, M. 2021: 'Variable kinship patterns in Neolithic Anatolia revealed by ancient genomes' *Current Biology*. https://doi.org/10.1016/j.cub.2021.03.050

Zeder, M.A. 2008: 'Domestication and early agriculture in the Mediterranean Basin: Origins, diffusion, and impact' *Proceedings of the National Academy of Sciences of the United States of America* 105.33: 11597–604. https://doi.org/10.1073/pnas.0801317105

8. The effects of agriculture on the human body: a case study of Anatolian Neolithic populations

Yılmaz Selim Erdal

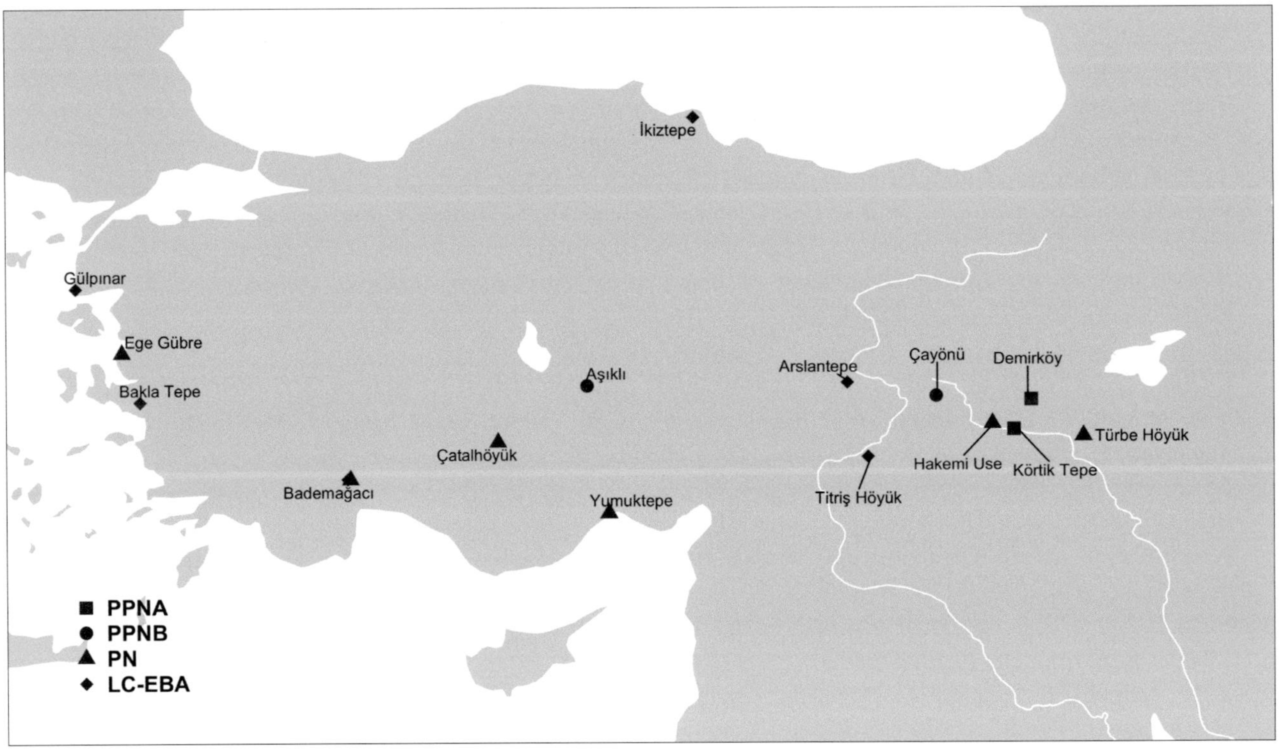

Fig. 8.1. Map showing the prehistoric sites in Anatolia examined in this study (by Nefize Ezgi Altınışık).

Introduction

The human skeleton has a plastic structure which is affected by physical and physiological stresses during ontogenesis. Studies on human populations suggest that the radical changes that occurred as part of the lifestyles of past people are reflected in their bones. Numerous investigations show that the size, shape and pathologies of bones vary according to the type and extent of activity performed in life (Angel 1984; Kennedy 1984; Frayer, Wolpoff 1985; Miller 1985; Trinkaus et al. 1994). Studies on dental remains propose that dental diseases such as caries and abrasion are related to an increased carbohydrate intake, and also to changes in food preparation techniques (Cohen, Armelagos 1984; Molleson 1994; 2000; Eshed et al. 2006). In several settings, skeletal changes include reduced stature and body mass, and other behaviour-related skeletal adaptations reflecting declining mobility (Angel 1984; Eshed et al. 2006; Mummert et al. 2011; Rosenstock et al. 2019).

The long-term transition from hunting-gathering-fishing to agriculture and herding in Anatolia started at around 10,500 BC and was accompanied by increased sedentism. With the Neolithic transition, funerary behaviours and rituals developed in terms of variety and complexity (Santana et al. 2015). Together with this shift in the system of behaviours and beliefs, the transition to different subsistence strategies also had different types of effects on the way of life, such as changes in dietary habits, exploited food resources, production-consumption patterns, daily activities, mobility, processes of craft manufacturing, specialisation of labour, social differentiation, and household and settlement patterns (Molleson 1994; 2000; Belfer-Cohen Bar-Yosef 2000; Byrd 2005; Eshed et al. 2006; Özdoğan 2014; Whittle 2014; Sagona, Zimansky 2015).

Other significant social and economic changes occurred during the Early Bronze Age (EBA) (Frangipane 2002; 2017; Şahoğlu 2005; Çevik 2007; Sagona, Zimansky 2015; Rothman 2017). One of the most important changes in arable agriculture is the introduction of irrigation systems in Mesopotamia and later in Anatolia. Moreover, in this period, agricultural production and consumption were mainly based on cultivated domestic crops. The presence of extensive storage systems suggests an intensification in crop production and the presence of agricultural surplus. The use of the plow (pulled by cattle) has been suggested for this period, and is provided as a further example of the changes in productivity intensification that agriculture and cultivation systems went through in the EBA (Frangipane 2002; Öz 2012). Irvine (Irvine et al. 2019; Irvine, this volume) has also proposed that dietary habits narrowed in the EBA period, and an *EBA dietary package* is present that was based on staple domestic agricultural products. All of this evidence suggests a move towards agricultural mass production and changes in the production systems of agricultural goods. Moreover, complex structures, including surplus storage units and monumental administrative, religious and ritual buildings developed. Settlements started to also be surrounded by fortification walls (Matney, Algaze 1995; Frangipane 2002; Laneri 2007). The rise of a central authority together with the emergence of elite groups was the basis for the formation of a complex and stratified society. Mineral resources, as well as production-consumption processes, experienced some of the biggest transformations and arguably played the most important role in shaping aspects of prestige, status and social structure. Metal procurement, for the production of both weapons and other objects as indicators of social status led to the development of long-distance trade and exchange networks (Frangipane 2002; Şahoğlu 2005; Efe, Fidan 2006; Çevik 2007; Rothman 2017). This increased connectivity likely encouraged population mobility that, in turn, affected genetic relationships even between communities geographically distant from each other, resulting in increased population diversity (Skourtanioti et al. 2020). All of these cultural and societal changes may also have had an effect on the human body in relation to, and as part of, the changing practices of production of agricultural items.

Paleopathology at the Origin of Agriculture is one of the pioneering studies that deals with the effects of cultural changes on population structure, health and body composition globally during the transition to agriculture (Cohen, Armelagos 1984). It is acknowledged that agricultural lifeways changed the types of mechanical loading on the bones. Anthropological studies suggest that osteoarthritis of the spine and of the main joints are diagnostic skeletal changes caused by increased physical activity and stress (Goodman et al. 1984; Molleson 1994; 2000). Human bones are dynamic structures formed by processes of absorption and remodelling. According to Wolff's law, bone has plasticity, and is sensitive and adaptable to mechanical and/or physical forces. It has been demonstrated that in individuals performing intense physical activity, bones respond to the intensity, frequency and direction of the physical stress by changing its density, volume and shape (Nilsson, Westlin 1971; Jones et al. 1977; Kennedy 1983; Larsen 1997; Knüsel 2000). Together with osteoarthritis, the external dimensions and cortical thickness of bones are also taken into consideration for better understanding and interpreting skeletal mechanical loading processes (Bridges 1989; Knüssel 2000; Ruff 2005; Auerbach, Ruff 2006). Bony ridges as the attachment sites for the anchoring of muscles, tendons and ligaments, as well as other bony reliefs of various sizes and shapes, are among the most important elements examined when analysing physical strength and human activity (Dutour 1986; Kennedy 1989; Hawkey, Merbs 1995; Steen, Lane 1998; Eshed et al. 2004).

Studies on prehistoric skeletal human remains conclude that mechanical loading on the human body increased with the transition to agriculture, as well as sex differentiation in labour division related to changes in subsistence strategies. However, studies on prehistoric skeletal human remains have not yet reached a general consensus on the exact and overall effects that the Neolithic transition had on the human body, and on gender differentiation in labour divisions related to the changes in subsistence strategies. In fact, while some studies state that the transition to agriculture increased the workload of male individuals (Goodman et al. 1984; Bridges 1991; Shuler et al. 2012), others have instead proposed an increase in the workload of females (Eshed et al. 2004). Bridges (1989), who examined the effects of agriculture on the cortical thickness of long-bone diaphyses, concluded that the workload increased for both sexes, albeit affecting different parts of the body. Molleson (1994) confirmed this result for Mesopotamian prehistoric people. Peterson (1997) proposed a more distinct division of labour by sex in hunter-foragers while suggesting that an increased workload in the Neolithic actually reduced the observed differences between the sexes.

Anthropologists have interpreted the effects of the Neolithic transition on the human body by analysing human skeletal remains, mostly from the Levant and

Northern Mesopotamia (Molleson 1994; 2000; Peterson 1997; Eshed et al. 2003; 2004; 2006; Smith, Horwitz 2007). With the advent of the Neolithic period, changes in subsistence strategies were also experienced in Anatolia. However, this geographical region was not considered to be relevant until the 1960s, after which archaeological studies demonstrated that both the number of settlements and human burials increased dramatically with sedentism. However, very few diachronic studies exist which examine the effects that changes in subsistence strategies have on populations in Anatolia. Angel (1984), Güleç (1989) and Koca-Özer et al. (2011) have all suggested a decrease in body height in the whole of the Mediterranean region from the adoption of agriculture onwards. Angel (1984) also proposed changes in the frequencies and incidences of anaemia and osteoarthritis. Y.S. Erdal (1999) discovered a 16% reduction in the survival of the total dental crown in Anatolia from the Neolithic onwards. Stable isotope studies which have been performed on Anatolian populations living between the Neolithic and the EBA indicate that dietary habits were based on a broad (isotopically) spectrum of food sources that narrowed with a move towards extensive arable agriculture and economically intensive animal husbandry during the EBA (see Irvine, this volume; Irvine et al., forthcoming, and references within both). Itahashi et al. (2019) proposed that the exploitation of freshwater fish resources decreased during the Pottery Neolithic (PN) in southeastern Anatolia. Analysis of auditory exostosis confirmed the decrease of aquatic activities in Anatolia after the PrePottery Neolithic (PPN) (Özbek 2012; Koruyucu et al. 2018). Furthermore, according to many scholars (Koca-Özer et al. 2006; Yılmaz, Pehlevan 2015; Koruyucu in this volume), the incidence and frequency of dental caries increased with the adoption of agriculture.

The main goal of this study is to examine the effects of an agricultural lifestyle on Anatolian populations, at both the physical and social levels. To that end, degenerative bony changes such as osteoarthritis, external dimensions and entheseal changes of the upper extremities, as well as aberrant wear on the anterior dentition, are investigated. Changes in sex differentiation concerning the adoption of different subsistence strategies are also investigated.

Materials and methods

Human remains studied within the scope of this study are dated to a wide time span, from the PPNA (ca 10,500 BC) to the end of the EBA (ca 2200 BC) (Fig. 8.1). The PPNA periods are represented by the permanent settlements of Körtik Tepe and Demirköy, both located in the Diyarbakır province of southeastern Anatolia and occupied by hunter-gatherer-fishermen (Özkaya, Coşkun 2007; Özkaya 2009; Benz et al. 2015; 2017). PPNB settlements include Çayönü (Özdoğan, Özdoğan 1989) also in Diyarbakır and Aşıklı Höyük in Aksaray, central Anatolia (Özbaşaran et al. 2018). Settlements dated to the Pottery Neolithic (PN) are Hakemi Use in Diyarbakır (Tekin 2011; 2013; Y.S. Erdal 2013) and Türbe Höyük in Siirt (Kodaş et al. 2018), both in southeastern Anatolia; Yumuktepe in Mersin (Y.S. Erdal 2019b) and Bademağacı in Antalya (Duru, Umurtak 2019), both in southern Anatolia, and Ege Gübre in Izmir, located in western Anatolia. Except for Hakemi Use, few individuals were recovered from the PN settlements. Gülpınar in Çanakkale in northwestern Anatolia (Erdal, Takaoğlu 2021), Arslantepe in Malatya (Frangipane 2002; 2017) in eastern Anatolia and İkiztepe in Samsun in the central Black Sea region (Bilgi 1984; 1990; Welton 2010) are dated to the Late Chalcolithic (LC). EBA sites include Titriş Höyük in Şanlıurfa, Bakla Tepe in Izmir and Bademağacı in Antalya (Algaze, Matney 2011; Duru, Umurtak 2012; Gündoğan et al. 2019); see Figure 8.1 for the location of all of these sites.

Published and unpublished data acquired by the analyses conducted on the human skeletal remains currently housed in the Husbio_L Laboratory of Hacettepe University were examined for this study. Sex and age-at-death were estimated using standard methods (Buikstra, Ubelaker 1994). Osteoarthritis was mainly investigated by Ö.D. Erdal (2004), and by the author of the present work, especially for samples from the more recently excavated sites. Entheseal changes were analysed according to the methods proposed by Hawkey and Merbs (1995). Data collected by Demet Delibaş (2016) as part of her MA thesis have also been included here. Medial epicondylitis was analysed in order to understand physical stresses on the elbow, the presence of porosity and the development of bony spicules for the development of epicondylitis (Spigelman et al. 2012), while dental abrasion was considered in terms of using the teeth as a tool (Y.S. Erdal 2008). Notches, grooves and lingual surface attrition on maxillary anterior teeth (LSAMAT) were also used to understand human activities (Y.S. Erdal 2021).

Results

Osteoarthritis as an indication of workload

Joints, which have different biomechanical properties, are affected in different ways by mechanical stresses as a result of habitual cultural behaviours (Jurmain 1977a; 1977b; Bridges 1991; Molleson 1994). Although many reasons such as trauma, hereditary diseases, immune disorders and/or infectious diseases may affect the joints,

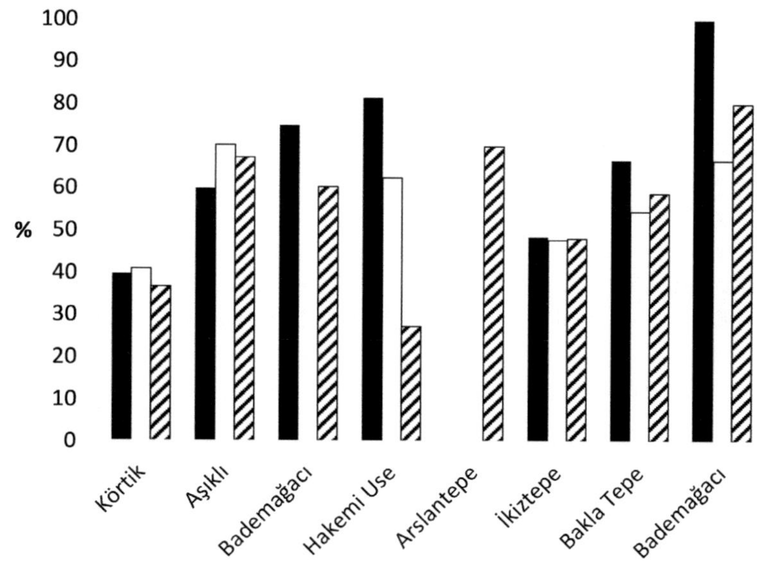

Fig. 8.2. Osteoarthritis rates from different prehistoric settlements in Anatolia (by Y.S. Erdal).

the repeated and continuous load of physical activity are accepted as the most common aetiological factors (Steinbock 1976; Larsen 1997; Waldron 1997). Degenerative joint diseases play an important role in determining physical stresses and understanding the different types of movement, activities and occupations of ancient human populations (Goodman et al. 1984; Jurmain 1990; Larsen 1997; Ö.D. Erdal 2004; Eshed et al. 2010). The frequency of degenerative lesions, their skeletal distribution and severity are considered fundamental for reconstructing the lifestyle or behavioural patterns of past communities (Steinbock 1976; Merbs 1983; Rathbun 1984; Larsen 1997; Ortner 2003). In this study, degenerative changes on the upper and lower extremities, rather than the spinal column, are used to interpret occupational activities and workload.

The data from different Neolithic Anatolian populations demonstrate an increase in the frequency of osteoarthritis from the PPNA to PN (Fig. 8.2). Despite the diachronic increase in frequency, the distinct sites show differences, both in terms of population and by sex. Combined data from different periods such as the PPNA, PPNB, PN and LC-EBA have been used to outline diachronic changes in osteoarthritis. Figure 8.3 shows an increase from the PPNA to PN and a decrease from the PN to EBA. Frequencies in the LC-EBA decreased almost to the level of populations characterised by a foraging lifestyle.

Differences between the sexes in degenerative joint diseases were observed in all chronological periods. During the PPNA, although both sexes have a similar ratio of osteoarthritis, females have a slightly higher (but

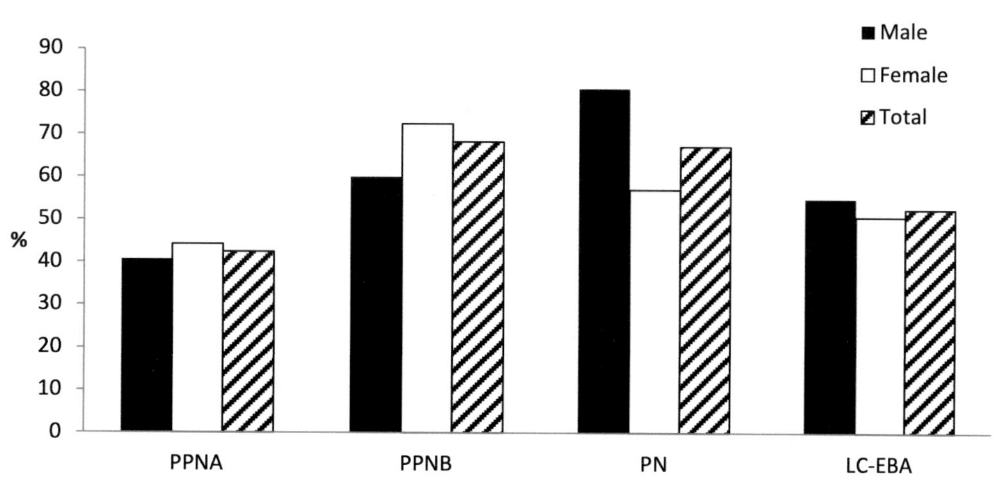

Fig. 8.3. Osteoarthritis rates by different time period (by Y.S. Erdal).

statistically insignificant) frequency ($p<0.005$) that increases with the transition to the PPNB (Fig. 8.3). Male individuals instead have a prominent and regular increase from the PPNA to PN. Differences in frequency between the sexes (approximately 12%) suggest a clear sex differentiation in early arable and livestock agricultural people of the PPNB. Even more evident differences can be observed for the PN people. A difference of nearly 20% is observed, with males having higher frequencies. These findings suggest not only an important increase in osteoarthritis from hunting-foraging to agrarian societies, but also an increase in sex-related differences following the adoption of agrarian lifestyles. Even if male individuals have a higher frequency during the EBA, a general decrease is seen in both sexes relative to earlier agricultural time periods (Fig. 8.3).

The external size and shape of bones

Bone dimensions

Longitudinal and circumferential dimensions of the major long bones were studied as part of investigating the physical response to environmental stimuli. The longitudinal dimension of bone has a high heritability, although it is also affected by dietary habits (Susanne 1977; Wood et al. 2014). Dietary habits, hygiene and socioeconomic status may all have positive or negative effects on bone length (Larsen 1995; 1997; Bogin 2001).

Maximal Length	PPNA			PPNB			LN			LC-EBA		
Males	*n*	*X*	*SD*	*n*	*X*	*SD*	*n*	*X*	*SD*	*n*	*X*	*SD*
Humerus R	6	322.0	23.45	2	296.0	5.66	3	308.0	13.26	18	307.3	19.31
Humerus L	6	316.6	13.77	2	291.0	9.90	1	293.5	.	21	307.0	19.73
Radius R	18	250.5	17.49	2	237.5	7.07	3	225.5	14.31	20	235.2	16.61
Radius L	10	241.4	12.43	0			5	240.4	4.84	13	232.8	9.11
Ulna R	13	266.2	15.47	3	260.8	3.88	5	257.2	14.12	16	260.5	34.60
Ulna L	21	271.1	14.74	3	259.3	2.08	3	251.5	17.90	18	256.2	17.64
Females	*n*	*X*	*SD*	*n*	*X*	*SD*	*n*	*X*	*SD*	*n*	*X*	*SD*
Humerus R	12	291.2	26.89	3	301.3	17.56	4	278.4	20.63	26	287.9	16.09
Humerus L	17	293.9	13.37	3	284.0	16.93	2	263.5	31.82	22	286.6	14.75
Radius R	21	232.3	15.69	3	229.5	15.06	2	216.0	24.04	30	216.0	14.24
Radius L	12	222.4	14.52	6	228.3	11.41	1	220.5	.	30	212.9	11.59
Ulna R	21	251.3	12.32	5	249.1	12.71	6	226.4	10.77	16	236.9	13.03
Ulna L	17	250.4	14.77	3	249.3	14.36	8	221.7	26.04	21	229.7	13.25

Tab. 8.1. Maximal Length of the upper extremities in Anatolian populations from the PPNA to LC-EBA.

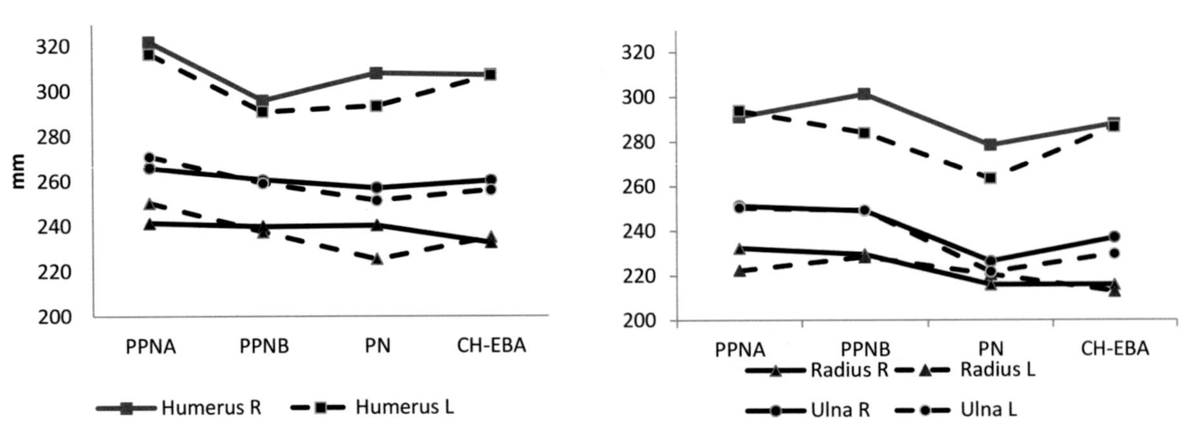

Fig. 8.4. Arm bone lengths from the Neolithic to the LC-EBA (by Y.S. Erdal).

However, transversal measurements, especially of the upper appendicular skeleton, provide insight into physical activities (Bridges 1989; Larsen 1997; Knüsel 2000; Auerbach, Ruff 2006), as demonstrated not only by investigations on ancient human remains (Ruff, Jones 1981; Bridges 1989; Trinkaus et al. 1994; Larsen 1997; Knüsel 2000) but also by more recent studies conducted on athletes and workmen performing intense physical activities (Duyar, Özener 2003).

There is a clear decrease from the PPNA to PPNB in the length of arm bones and a slight increase from the PPNB to LC-EBA (Tab. 8.1; Fig. 8.4). An increase in measurements of the radius and ulna is also seen from the PN to the LC-EBA period. A slight difference between the PPNA and PPNB is observed in females, with a sudden decrease in the length of all forearm bones. A slight increase is, however, seen in the humerus and ulna. Conversely, Arslan (2005) suggested a decrease in bone lengths from the Neolithic to EBA. When analysing the results in longitudinal measurements of long bones it is important to take into account differences between the sides. The long bones of the upper extremities show asymmetry with higher values on the right side (Fig. 8.3). Compared to forearm bones, differences between the sides are more obvious in the humeri of both sexes. Differences in asymmetry values are also diachronically visible. Both sexes of the PN period have higher asymmetry compared to the PPNA and LC-EBA. Additionally, females have higher humeral asymmetry in intensive agricultural societies, while males have side asymmetry in all arm bones.

A different situation has been observed with regard to bone circumferences. Circumferences of the upper extremities show an increase over time (Tab. 8.2). Compared to the early phases of the Neolithic, a clear increase is seen from the PN to LC-EBA. Similar to bone length, differences between the sides demonstrate higher values in right arm bones. The main differences are of the upper arm bones of male individuals, especially during the agricultural Neolithic period. Arslan (2005) found that the length, diameter and circumference of long bones increased from the Neolithic onwards. Conversely, robustness values decreased from the PN people to the more 'complex' societies of the LC-EBA period (Arslan 2005: 193–94). Asymmetry is more prominent on the upper extremities, and is higher in male individuals (Arslan 2005: 146). In agreement with Arslan (2005), I suggest that the length of long bones decreased while perimeters and diameters increased from hunter-forager to agricultural and intensive agricultural lifeways. These alterations imply a stocky shape of the bones, especially the humeri (Fig. 8.5). Conversely, during the LC-EBA, the robusticity and side asymmetry of these bones decreases. However, in this study, the robustness values of the right humeri suggest an increase from the PPNA onwards. Despite a slight increase, the PPNA and PPNB values are similar. A more relevant increase is instead seen from the PPNB to PN period, while a decrease occurs from the PN to the LC-EBA period. Compared to females, males have higher robustness values for both the humerus and radius (Fig. 8.5).

Circumferences	*PPNA*			*PPNB*			*LN*			*LC-EBA*		
Males	*n*	*X*	*SD*	*n*	*X*	*SD*	*n*	*X*	*SD*	*n*	*X*	*SD*
Humerus R	48	64.1	4.33	8	65.6	3.25	15	64.7	3.76	30	68.2	4.56
Humerus L	51	62.3	3.62	5	62.7	2.73	13	61.9	3.17	29	66.7	4.58
Radius R	43	40.5	2.75	3	39.3	2.26	16	40.2	2.11	22	43.3	2.68
Radius L	52	40.7	2.82	4	40.1	4.97	16	41.5	3.24	20	42.73	2.63
Ulna R	31	34.8	2.56	3	34.3	1.53	11	36.2	2.47	16	37.2	3.90
Ulna L	36	36.1	2.10	5	33.0	1.28	8	35.8	2.73	20	36.5	2.94
Females	*n*	*X*	*SD*	*n*	*X*	*SD*	*n*	*X*	*SD*	*n*	*X*	*SD*
Humerus R	56	15.5	1.65	17	15.5	1.26	24	15.3	1.76	44	15.9	1.70
Humerus L	57	56.2	3.31	10	55.6	3.48	25	55.8	4.19	46	59.2	4.02
Radius R	55	36.4	2.37	11	36.7	2.69	19	36.2	2.67	38	38.1	2.86
Radius L	51	36.4	2.77	8	36.6	1.51	21	36.9	2.76	44	37.8	2.59
Ulna R	35	32.9	1.98	11	32.7	1.92	16	32.4	2.16	28	32.6	3.63
Ulna L	33	32.6	2.16	6	31.5	2.68	18	31.5	2.68	26	32.7	2.15

Tab. 8.2. Circumferences of upper extremities in Anatolian populations from the PPNA to LC-EBA

Epicondylitis

From clinical studies, it is well known that forceful and repetitive manual activities and prolonged tension affect the forearms and elbows, and increase the risk of epicondylitis (Priest et al. 1977; Tschantz, Meine 1993; Marklin, Monroe 1998; Pascarelli, Hsu 2001; Walker-Bone et al. 2004). Stresses related to pronation, rotation, flexion and forearm extension, caused by micro-ruptures of the tendons in the epicondyles, ultimately result in degenerative tendinopathy in the form of entheses and bone pits or pores (Spigelman et al. 2012; Mark et al. 2015). Medial epicondylitis (golfer's elbow) occurs at the common flexor origin and is affected by repeated stress on the elbow and wrist, which is required for a strong flexion of the hand and fingers (Siri et al. 2006; Spigelman et al. 2012; Mark et al. 2015). The main reasons for epicondylitis are occupational activity/activities. This is why epicondylitis can be used as a marker of activity in past people, and to understand occupation, changes in lifestyle, sex-related differences and socioeconomic division of labour, as well as to reconstruct specific activities (Miller 1985; Crianni, Fornaciari 2003; Spigelman et al. 2012; Üstündağ 2020).

The data for medial epicondylitis suggest that its frequency is low in the PPNA period (nearly 5%), and then steadily increases throughout the PPNB and PN (Tab. 8.3). Conversely, a sudden decrease is observed in people of the LC and EBA. Males and females have different frequencies of medial epicondylitis. Differences between the sexes are very low (1%) during the PPNA. However, the frequency in males suddenly increases during the PPNB. While a slight increase is noted in the PPNB, a larger jump is observed in females during the PN (Tab. 8.4). The LC-EBA people have very low frequencies, resembling those observed in PPNA people whose subsistence strategies are characterised by hunter-forager lifeways. Even though differences were observed between the settlements dated to the same period (Fig. 8.6), the overall findings suggest a similar trend for Anatolia.

Aberrant wear on anterior teeth

Teeth lose their main tissues in different ways, such as through attrition, abrasion and abfraction. Tooth-tooth and/or tooth-food contact are two of the main reasons for dental tissue loss due to mechanical stress. However, biting, wetting, chewing or gripping non-food materials during production and processing may also cause abrasion (Mahoney, Kilpatrick 2003; Y.S. Erdal 2008; Irvine et al. 2014; Tomczyk, Zalewska 2016; Watson, Haas 2017; Crane et al. 2020). Abfraction and interproximal grooves are mostly seen on the posterior dentition, while grooves and notches are observed on the anterior teeth, which are frequently used for biting, holding, scraping, sucking or wetting materials during production processes (Y.S. Erdal 2008). Similar to the previously discussed bony changes and dental abrasion, examination of the ways in which people used their teeth as a tool has been used to reconstruct occupational activities and production techniques and processes, as well as understanding the daily activities of past people, changes in lifeways, sex, age, and socioeconomic intragroup differences and intergroup differences.

All of the analysed prehistoric Anatolian sites are characterised by the presence of occupational activity-related abrasions on the anterior teeth (Tab. 8.5). The PPNA site of Körtik Tepe has the highest frequencies of dental grooves on incisors, as well as of notches and LSAMAT. Dental grooves and LSAMAT are also observed in the PPNB populations of Çayönü and

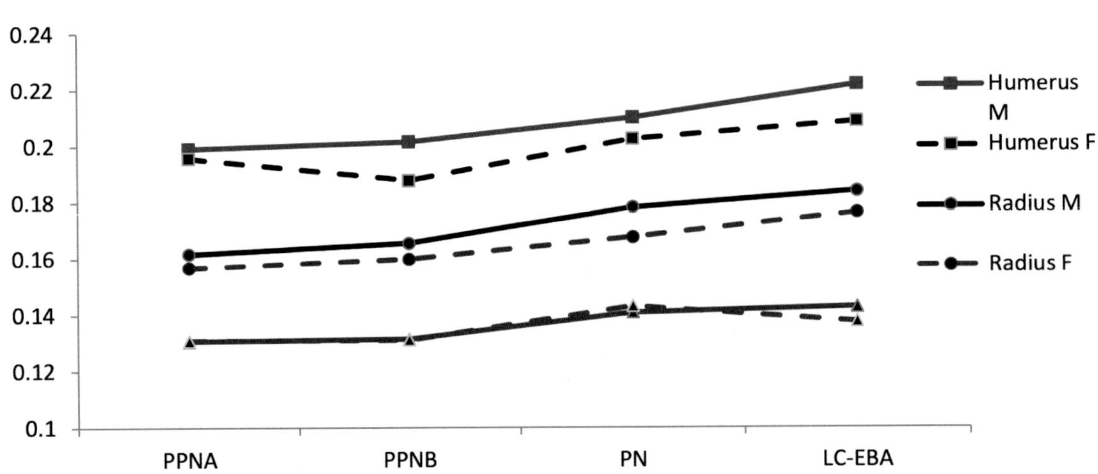

Fig. 8.5. Robustness values of arm bones from the Neolithic to EBA (by Y.S. Erdal).

Settlement	*Males*		*Females*		*Total*		χ^2
	N	*%*	*N*	*%*	*N*	*%*	*df:1*
Körtik Tepe	4/76	5.3	3/68	4.4	7/151	4.6	0.056
Demirköy	–	–	0/2	0.0	0/2	0.0	–
Aşıklı	2/10	20.0	1/16	6.3	3/26	11.5	12.140
Hakemi Use	1/9	11.1	4/20	20.0	5/29	17.2	0.344
Yumuktepe	1/1	100.0	1/7	14.3	2/6	25.0	3.429
Arslantepe	2/9	22.2	4/15	26.7	6/25	24.0	0.163
İkiztepe	0/22	0.0	0/33	0.0	0/55	0.0	–
Bademağacı	0/2	0.0	0/2	0.0	0/4	0.0	–
Bakla Tepe	0/6	0.0	0/12	0.0	0/18	0.0	–
Acemhöyük	1/2	50.0	0/7	0.0	1/8	11.1	3.938
Titriş Höyük	1/3	33.3	0/1	0.0	1/4	25.0	0.759

Tab. 8.3. Frequency of medial epicondylitis in prehistoric Anatolian populations

Period	*Males*		*Females*		*Total*		χ^2
	O/A	*%*	*O/A*	*%*	*O/A*	*%*	*df:1*
PPNA	4/75	5.3	3/70	4.3	7/151	4.6	0.075
PPNB	2/10	20.0	1/16	6.3	3/26	11.5	0.277
PN	1/8	12.5	5/20	25.0	6/28	21.4	0.329
LC-EBA	4/44	9.1	4/70	5.7	9/114	7.0	0.472

Tab. 8.4. Changes in the frequency of medial epicondylitis

Aşıklı. At Çayönü, dental grooves are observed in males, albeit at low frequencies. At Aşıklı, only one example was found in males. The PN site of Hakemi Use is one of the prehistoric Anatolian sites with the highest frequencies of dental grooves, and these grooves are found exclusively in males. The variety and the frequencies of occupational abrasion at Hakemi Use is comparable to those of the PPNA sites. At Yumuktepe the unique aberrant wear observed is LSAMAT. Among LC-EBA communities, Arslantepe is characterised by frequencies similar to those of the LN communities. Although the number of teeth is limited in this sample population, grooves and LSAMAT are observed in both sexes. No grooves are observed at İkiztepe, which is contemporary with Arslantepe, or at the EBA settlements of Bakla Tepe and Titriş Höyük. At these sites, apart from a few notches observed in females, dental wear does not seem to be related to the use of anterior teeth as pinchers. Conversely, LSAMAT is the most common aberrant wear. Despite differences between communities, a general decrease in aberrant wear is observed over time.

Figure 8.7 clearly shows that LSAMAT is present in all time periods and in both sexes. There are no statistical differences between the chronological periods. Notches are also seen in almost all periods, with the exception of the PPNB and the very few cases from the LC-EBA. Özbek (2000) found a notch on an isolated central incisor from Öküzini, and notches were observed on three teeth from the Late Chalcolithic settlement of Çamlıbel Tarlası (Irvine et al. 2014). The most important changes are observed with regard to occlusal grooves, which are common in the PPNA period but decrease from the PPNA to PPNB. The PN people have higher frequencies of occlusal grooves, while the LC-EBA people have few grooves. It is clear that the frequency of occlusal grooves decreases over time, and the differences between cultural horizons are statistically significant. Other relevant changes are the increased differences between sexes. Although frequencies are similar in both sexes during the PPNA period and they decrease over time, occlusal grooves are observed in all prehistoric periods. Nevertheless, a different pattern is seen in males. A clear decrease in the frequency of occlusal grooves is observed

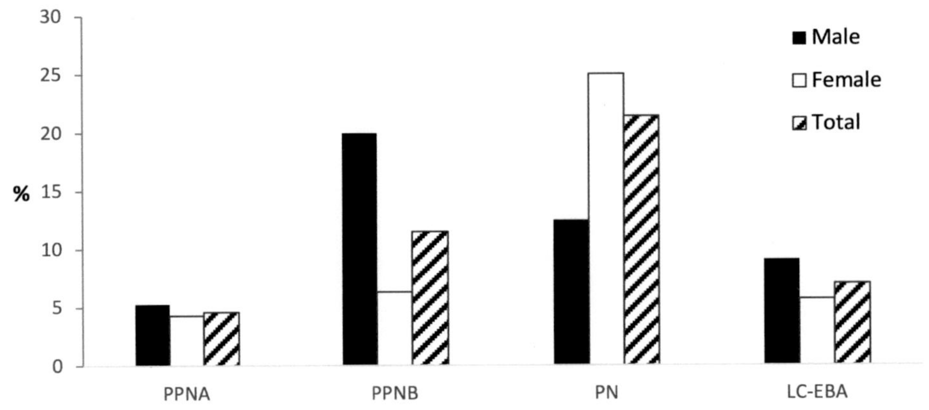

Fig. 8.6. Rates of medial epicondylitis in prehistoric Anatolian populations (by Y.S. Erdal).

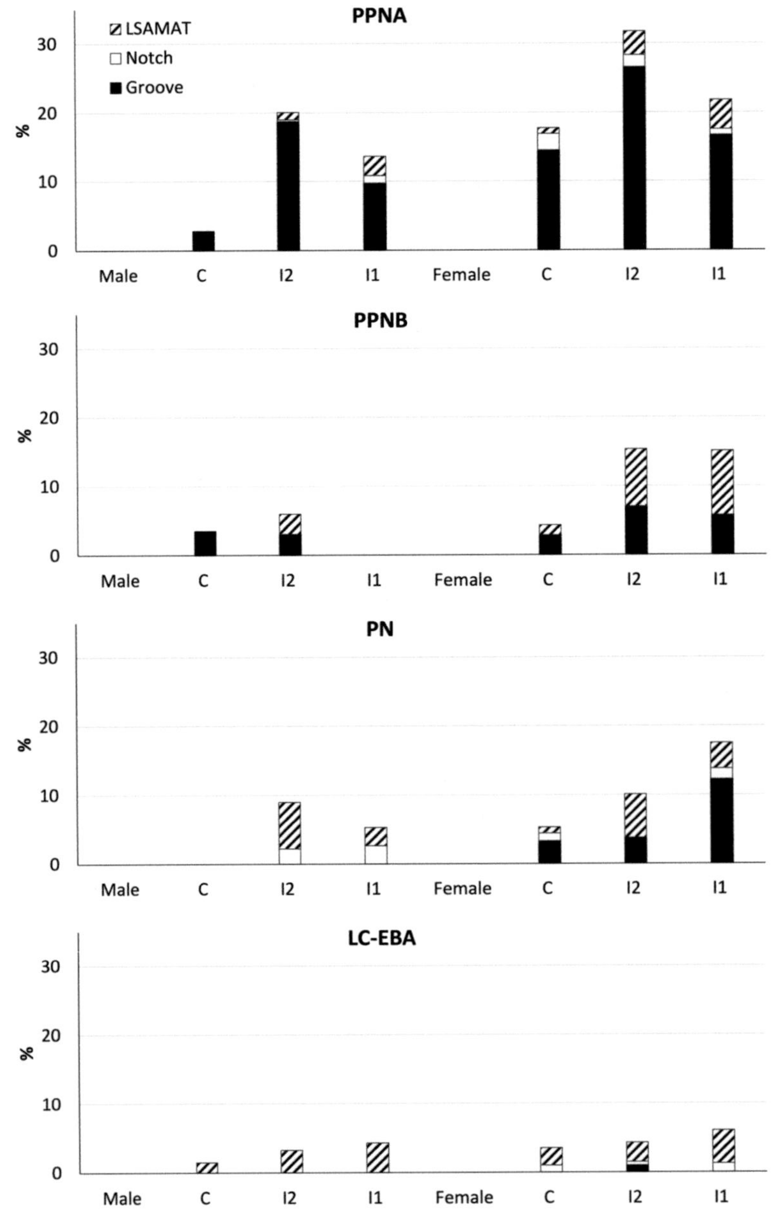

Fig. 8.7. Aberrant dental wear in prehistoric Anatolian populations (by Y.S. Erdal).

			I1		*I2*		*C*	
			N	*%*	*N*	*%*	*N*	*%*
KTK	♂	N	156		134		189	
PPNA		G	15	9.6	25	18.7	5	2.6
		No	2	1.3	1	0.7	-	-
		L	4	2.6	1	0.7	-	-
	♀	N	115		110		125	
		G	19	16.5	29	26.4	18	14.4
		No	1	0.9	2	1.8	3	2.4
		L	5	4.4	4	3.6	1	0.8
ÇT	♂	N	9		13		11	
PPNB		G	-	-	1	7.7	-	-
		No	-	-	-	-	-	-
		L	-	-	1	7.7	-	-
	♀	N	18		19		19	
		G	2	11.1	2	10.5	1	5.2
		No	-	-	-	-	-	-
		L	2	11.1	1	5.3	-	-
AH	♂	N	19		20		23	
EN		G	-	-	-	-	-	-
PPNB		No	-	-	-	-	-	-
		L	-	-	-	-	-	-
	♀	N	35		41		51	
		G	1	2.9	2	4.9	1	1.9
		No	-	-	-	-	-	-
		L	3	8.6	4	9.8	1	1.9
HKM	♂	N	32		37		45	
PN		G	-	-	-	-	-	-
		No	1	3.1	1	2.7	-	-
		L	1	3.1	1	2.7	-	-
	♀	N	37		55		60	
		G	7	18.9	3	5.5	3	5.0
		No	1	2.7	-	-	1	1.6
		L	1	2.7	2	3.6	1	1.6
YTP	♂	N	5		7		7	
PN		G	-	-	-	-	-	-
		No	-	-	-	-	-	-
		L	-	-	2	28.6	-	-
	♀	N	20		23		28	
		G	-	-	-	-	-	-
		No	-	-	-	-	-	-
		L	1	5.0	3	13.0	-	-
ARS	♂	N	9		4		10	
LC		G	1	11.1	-	-	-	-
EBA		No	-	-	-	-	-	-
		L	1	11.1	-	-	-	-

Tab. 8.5 (cont. on next page).

			I1		*I2*		*C*	
			N	*%*	*N*	*%*	*N*	*%*
ARS	♀	N	16		15		19	
LC		G	-	-	2	13.3	1	5.3
EBA		No	-	-	-	-	-	-
		L	4	2.6	2	1.2	2	1.0
İT	♂	N	151		168		209	
LC		G	-	-	-	-	-	-
EBA		No	-	-	-	-	-	-
-		L	4	25.0	-	-	-	-
	♀	N	195		208		230	
		G	-	-	-	-	-	-
		No	3	1.5	-	-	-	-
		L	-	-	-	-	-	-
BT	♂	N	32		39		43	
EBA		G	-	-	-	-	-	-
		No	-	-	-	-	-	-
		L	2	6.3	3	7.7	1	2.3
	♀	N	51		58		63	
		G	-	-	-	-	-	-
		No	-	-	1	1.7	-	-
		L	8	15.7	7	12.1	8	12.7
TH	♂	N	35		34		43	
EBA		G	-	-	-	-	-	-
		No	-	-	-	-	-	-
		L	1		2		-	-
	♀	N	15		22		31	
		G	-	-	-	-	-	-
		No	-	-	-	-	-	-
		L	1	6.7	1	4.5	1	3.2

Tab. 8.5. Frequencies of abrasion on the anterior teeth of Anatolian populations. ***N:*** *number of teeth;* ***G:*** *occlusal groove;* ***No:*** *notch;* ***L:*** *LSAMAT;* ***AH:*** *Aşıklı Höyük;* ***ARS:*** *Arslantepe;* ***BT:*** *Bakla Tepe;* ***ÇT:*** *Çayönü Tepesi;* ***HKM:*** *Hakemi Use;* ***İT:*** *İkiztepe;* ***KTK:*** *Körtik Tepe;* ***TH:*** *Titriş Höyük;* ***YP:*** *Yumuktepe;* ***PPNA:*** *Pre-Pottery Neolithic Period A;* ***PPNB:*** *Pre-Pottery Neolithic Period B;* ***PN:*** *Pottery Neolithic Period;* ***LC:*** *Late Chalcolithic;* ***EBA:*** *Early Bronze Age.*

in males from the PPNA to PPNB, while dental grooves are not observed in males after the PPNB period (Fig. 8.7). This data suggests that the use of teeth as a tool was not performed by male individuals after the PPNB and/or they did not need to take part in production processes which required using the teeth as a tool.

Discussion

Temporal changes in workload

Studies on human health conditions suggest that morbidity increased with the advent of increased sedentary life during the shift from a hunting-foraging to an agricultural lifestyle (Cohen, Armelagos 1984; Arensburg 1985; Larsen et al. 2015). One of the most consistent findings in studies about the Neolithic transition is an evident decrease in body height (Larsen 1981; 1984; Ruff et al. 1984; Mummert et al. 2011). Larsen (1995), for instance, found that both males and females of prehistoric populations living on the Georgia Coast (USA) demonstrated decreased dimensions and strength of the long bones during the agricultural period, thus suggesting a reduction in workload compared to pre-agricultural times. Similarly, other investigations (Ruff et al. 1984; Bridges 1991; Ö.D. Erdal 2004) have proposed that physical stress decreased with the transition to agriculture. Conversely, Bridges (1989) and Larsen and Ruff (1994) suggested that physical activities and stress increased with the adoption of an agricultural lifestyle. The result of this physical stress is seen in the thicker and stronger long bone diaphyses of agriculturalists compared to hunter-foragers (Bridges 1983; 1985; 1989; Larsen 1997; Arslan 2005). A decrease in the length of long bones (Angel 1984) and an increase in circumferential dimension and robustness among Anatolian people (Aslan 2005) were interpreted as coinciding with the adoption of agriculture. According to Larsen (1997), bone cortical thickness was also affected. Studies on the skeletal population from Neolithic Çatalhöyük (Larsen et al. 2015) and European human skeletal remains (Marchi et al. 2006) have indicated an increase in bone mass. Analysis of cross-sectional geometry and external bone measurements of the Çatalhöyük people suggest that workload increased in the Middle Period, when agriculture and animal husbandry became widespread at the settlement (Larsen et al. 2015; 2019).

Investigations on entheseal changes have revealed changes in workload with the advent of agriculture. It is known that stress on the bone is essential for osteoblastic activity (Kennedy 1989; Hawkey, Merbs 1995; Steen, Lane 1998; Weiss 2007; 2009). Entheseal changes can also be used to reconstruct the daily life and occupational activities of prehistoric people (Dutour 1986; Molleson, Blondiaux 1994; Steen, Lane 1998; Lovell, Dublenko 1999; Molleson, Hodgson 2000; Eshed et al. 2004; Molnar 2010; Shuler et al. 2012). Eshed et al. (2004) examined Levantine Natufian foragers and Neolithic agricultural communities by analysing 21 muscle and ligament attachments of the upper limbs. They noticed that entheseal changes were more prominent in the Neolithic community than in hunter-foragers, and stated that the Neolithic transition resulted in higher workloads on the upper limbs. A greater workload among agriculturalists was also observed at Abu Hureyra, northern Mesopotamia (Molleson 1994). Similar to the Levant, studies in the New World have also produced similar results and conclusions (Bridges 1983; 1989; Goodman et al. 1984; Shuler et al. 2012).

Larsen et al. (2015) suggest that with the intensification of agriculture in the Middle Period of Çatalhöyük, the frequency of osteoarthritis was observed to increase. The most detailed study on osteoarthritis in Anatolian skeletal samples dated from the PPNB period up to more recent periods was conducted by Ö.D. Erdal (2004), who investigated the frequency of osteoarthritis and its degree of development in a total of 622 individuals from 14 sites in Anatolia. She found a decrease in frequency from the Early Neolithic (78.2%) to the EBA (69%). She also noticed that the shoulder, knee and elbow were the most affected joints among Neolithic people. The distribution and severity of the lesions were interpreted as the result of harsh physical stress related to hunting-gathering lifeways rather than agrarian (Ö.D. Erdal 2004). The Neolithic remains included in her study came from the PPNB sites of Çayönü and Aşıklı, and both had mixed subsistence strategies, where hunting-gathering was succeeded by small-scale agrarian lifeways (Erim-Özdoğan 2011; Düring 2013). It seems possible, therefore, that the decrease in frequency of osteoarthritis was not related to the introduction of intensive agriculture. The frequency of osteoarthritis during the EBA was observed mostly in the knee joints of individuals, followed by the shoulder and hip joints (Ö.D. Erdal 2004). Thus, it is clear that the frequency of osteoarthritis in the EBA follows a different pattern compared to the Early Neolithic period (Tab. 8.6).

Currently there have been no diachronic studies on entheseal changes in Anatolia. Delibaş (2016) analysed 178 adult individuals from Körtik Tepe who likely followed a hunter-forager lifestyle. It was acknowledged in that study that the muscular markers were not severe enough to recognise heavy entheseal lesions (Fig. 8.8; Delibaş 2016). She demonstrated that physical stress was

	Parameter	*Humeral robustness*	*Osteoarthritis*	*Entheseal changes*	*Medial epicondylitis*	*Occlusal groves*
PPNA	Thin	Low	Low	Low	Low	Very high
PPNB	Medium	Slightly increased	High	–	Medium	Medium
PN	Medium	Increased	High	–	High	Medium
LC-EBA	Thick	Increased	Medium	–	Low	Low

Tab. 8.6. Comparative results of the different skeletal and dental aspect in foragers and agrarian peoples of prehistoric Anatolia.

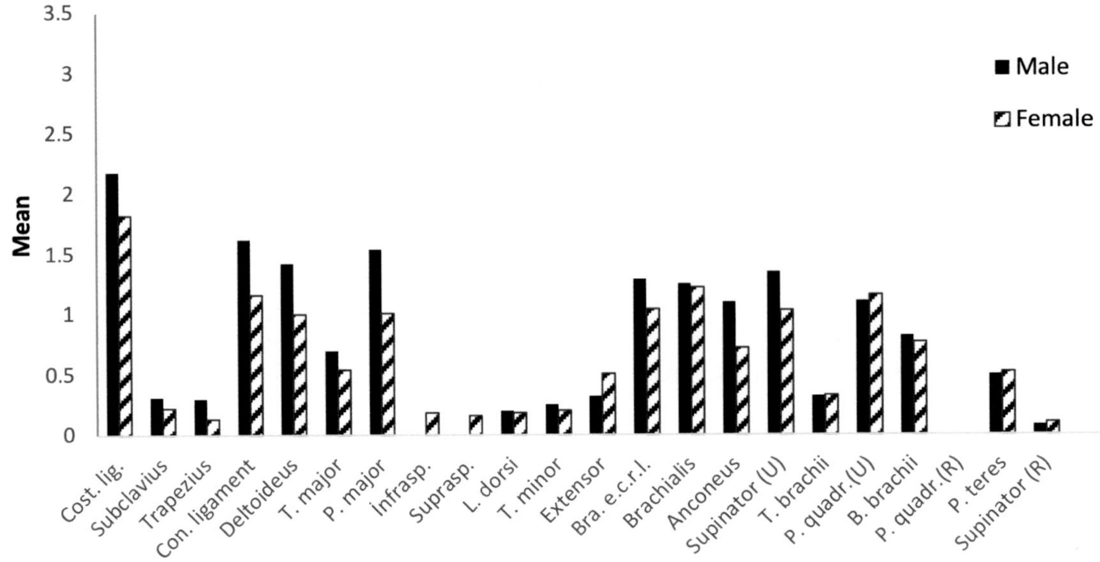

Fig. 8.8. Entheseal changes on arm bones at Körtik Tepe (after Delibaş 2016).

more severe on the arms than the legs. Similar to other ethereal changes, data on epicondylitis in this study show that the rate of medial epicondylitis is very low. Even if physical stress was not severe, based on the variations observed in the development of different muscle/tendon attachments in different long bones, and following archaeological data, Delibaş (2016) suggested that the Körtik Tepe people constructed small-sized houses, exploited terrestrial and freshwater resources, practised fishing, harvesting, milling and the grinding of grasses with seeds, as well as weaving nets and textiles. Entheseal changes showing greater development on the lower limbs are the result of activities such as walking short distances, running, jumping and climbing. Prominent entheseal changes on the shoulders were explained as the effect of activities such as carrying loads, using pestles, mortars and hand drills as grinding mills, all of which were demonstrated by the archaeological material culture finds to be very common at Körtik Tepe (Özkaya et al. 2010; Coşkun et al. 2011; Özkaya, Coşkun 2011). The epicondylitis data also fits with the archaeological data and enthesis.

Compared to other periods, skeletal data on body height, dietary habits and dentition suggest that the Late Epipalaeolithic and PPNA people at Körtik Tepe did not have to work hard and in an arduous manner due to the richness of their local environment and the abundance of resources at their immediate disposal. The huge amount of grave goods accompanying the Körtik Tepe burials (Özkaya, Coşkun 2011; Özkaya et al. 2012; Benz et al. 2016a), as well as the discovery of traces of red ochre and black dye scattered on the corpses, the loosely woven textiles used for wrapping them and the gypsum plasters (Coşkun et al. 2011; Özkaya, Coşkun 2011) reflect the importance that post-depositional treatments had in the system of beliefs and funerary rituals of that chronological period (Y.S. Erdal 2015). Anatolian, north Mesopotamian and Levantine Early Neolithic settlements seem to have lived in prosperity, or at least it seems that they did not need to spend time and work that required heavy physical stresses. Studies utilising stable isotopes, dental caries and body height in relation to dietary habits suggest that the Early Neolithic people (at least at Körtik Tepe) relied on a broad spectrum of food sources consisting of a variety of wild animals, fish, tortoise, wild legumes, cereals and fruits (Coşkun et al. 2011; Benz et al. 2016b). The archaeological and anthropological evidence, therefore, suggests that the Early Neolithic Körtik Tepe, and possibly all Early Neolithic Anatolian people, did not endure high levels of physical stress, and they had enough space and time to produce goods and perform complex rituals.

The highest frequencies of occlusal grooves, which were found in the Körtik Tepe people, suggest that they were engaged in the production of yarns, textiles, rugs, mats and baskets made from plant and animal fibres (Coşkun et al. 2011; Y.S. Erdal 2015; Koruyucu et al. 2018). Studies on the degree and patterns of abrasion of the anterior teeth among the earliest people across the world show similar aberrant wear patterns. Based on the present data, it is possible to state that the Körtik Tepe people have some of the highest frequencies of occupational dental abrasion in the world (Molleson 1994; Y.S. Erdal 2008). The remnants of textiles, fabrics, yarns and dyes possibly associated with such materials recovered from the skeletal remains of the PPNA settlements of Körtik Tepe, Hasankeyf Höyük and Demirköy suggest that producing these types of materials and artefacts was important in Anatolia. Entheseal changes observed in the brachialis, anxious, supinator and pronator quadratus muscle attachment centres might be related to this production as well. They might be caused by the unusual posture of the forearms and hands during the working process, thus suggesting an intensive use of the hands as tools (Walz et al. 2010; Mark et al. 2015).

All of the data regarding bone measurements, robustness, entheseal changes and medial epicondylitis show that the body composition and bone mass of the PPNA people of Anatolia were similar to those observed in Natufian hunter-foragers (Eshed et al. 2004), and that there were likely low levels of physical stress on the body compared to later peoples. It is well known that there was a continuity in hunting, gathering and fishing activities during the PPNB period. A broad spectrum of hunting sources, including the remains of a variety of small prey, birds and fish, as well as domestic animals, have been discovered at PPNB settlements (Özbek 2004; Erim-Özdoğan 2011; During 2013; Stiner et al. 2018). At Aşıklı, evidence of fences that were likely used for penning animals were found, as well as the bones of domestic sheep (Özbaşaran et al. 2018). Archaeobotanical studies suggest that wild cereals were an important food source during the Late Epipalaeolithic and PPNA periods (Riehl et al. 2011). However, together with wild plants, domesticated cereals were also important (Buitenhuis et al. 2018; Özbaşaran et al. 2018). All of these types of data suggest a mixed economy based on hunting-gathering activities supported by agropastoralism. These types of subsistence lifeways suggest that there would have been an increased variation of daily life that would have affected people over a long period of time. Digging the soil for small-scale agricultural activities might have increased the daily and seasonal workload of people in the PPNB. An important change occurred in the shift from using mortars to grinding stones to grind cereals (Bamyacı 2018). It has been suggested that some stone artefacts found at Natufian sites may be grinding stones. However, the use of grinding stones appears to have certainly increased in the PPNA period. At Körtik Tepe there are grinding stones, but these are mostly mortars. The use of saddle querns did not increase until after the full adoption of agriculture (Bamyacı 2018), and this is demonstrated by the large number of saddle querns found at PN settlements. It would not only have been agricultural activities such as tending to the fields and harvesting, etc., but also the construction of larger buildings for expanded families and structures for storing food that might have changed physical stress and stressors, as reflected by the increased entheseal changes observed on arm bones (Eshed et al. 2004). There is a wide variety of building structures from the PPNB period onwards (Özdoğan 2014). Digging the earth, cutting trees and bringing stones to construct more complex buildings with larger sizes and multiple rooms with basements, etc. requires a greater investment and amount of physical labour. Thus, together with slightly increased robusticity, decreased body height, increased osteoarthritis and epicondylitis, the PPNB people are probably physically representative of people between hunting-gathering and agricultural lifeways. Data from analysis of the bones supports an increased workload and physical stress, but the figures are not as severe as people who had fully adopted an agricultural way of life. Instead, they are positioned between hunter-foragers and fully agrarian people in Anatolia. Epicondylitis and robustness of the forearm, asymmetry on the arm bones and dental grooves also demonstrate frequencies and other patterns between PPNA and PN people. It has also been suggested that physical stress in PPNB lifeways did not increase suddenly, and did not affect every part of the body at the same time. In the PPNB, communities organised themselves in relation to new technologies, collective symbolism and ideas that increased the appeal of this new way of life (Duru 2015). However, this new way of life seems to have also increased the physical stresses of the early agriculturalists.

At the end of the PPNB period there is a collapse, with evidence for cultural changes in Anatolia and beyond (Duru 2015). The building of cultic centres, burial customs, the type of animal husbandry, the sizes of buildings, settlement patterns and the number of settlements all show changes. One of the main changes observed in the PN is a decisive expansion of the Neolithic way of life (Düring 2013). In the PN there is a shift in the subsistence strategies to one that is more-or-less fully based on domestic agriculture. This involved intensive mixed farming with the cultivation of domestic cereals and domestic animal husbandry. Animals also began to be exploited for their ante-mortem products such as milk, for example (Evershed et al. 2008). Subsistence strategies, production and consumption, and rituals and ideology in the PN were more centred around the house and household (Düring, Marciniak 2005). This style and strategy of production is also known as household production (Bernbeck, Nieuwenhuyse 2013). With the move to a fully domestic agricultural subsistence strategy there were big changes with regard to the techniques of obtaining and processing plant-based nutrients. Furthermore, at Hakemi Use (and indeed other PN settlements) archaeozoological and archaeological analyses have demonstrated that in the PN almost none of the animals were wild, and there is no evidence for bows and arrows or spears that may have been used for hunting activities (Tekin 2013). The archaeobotanical results suggest that the main cultivated crops include domesticated wheat, barley, peas, chickpeas and bitter vetch. Fishing activities also seem to have lost their importance. A slight decrease in protein sources might indicate a move from fishing or hunting wild animals to a greater dependence on agricultural production, particularly domestic plants and crops (Itahashi et al. 2019). The transition from hunting-gathering through a mixed subsistence strategy to the adoption and employment of fully domesticated agriculture likely had an effect on the physical demands and workload of people in the PN period.

Many agricultural crops require around one year of work, indicating an intensive and *longue durée* workload. Bridges (1989) suggested that agriculture and using a hoe are more labour intensive than hunting-gathering, and that agricultural life was physically arduous and time-consuming, and characterised by a higher level of mechanical loading. Even though the exact cultivation practices are not easy to interpret, all the data suggests that these changes must have affected the increased variety of the workload of the PN people, and this is reflected in the increased frequencies of osteoarthritis, the increased robusticity and asymmetry. Studies on handedness have also suggested an increased asymmetry due to heavy physical activities and manual work (Özener 2007; 2010). The handedness, increased asymmetry and osteoarthritis might be the result of intensified agriculture and a change in production practices during the Late Neolithic. An important change also occurs in the shift from the use of mortars to grinding stones, and also the presence of husking trays used for harvesting and grinding cereals (Tekin 2013; Bamyacı 2018). The production of handmade pottery was also a new cultural element in this period, and was likely an occupational activity in the daily life of the LN people. From the PPNB onwards, asymmetry in the arm bones increased, especially on the humeri. This suggests that right hands were repetitively and frequently used during production activities. Almost all of the analysed aspects in this study suggest that there was a clear increase during intensive agriculture. This conclusion has also been suggested by many archaeologists working on different aspects of human remains (Molleson 1987; 1994; Eshed et al. 2004; Marchi et al. 2006; Pitre et al. 2017).

Compared to other skeletal pathologies, the use of teeth as a tool evidently decreased. With the advent of the agropastoralist lifestyle in the PN, the possible development and enrichment of the toolset and changes in production technologies may explain the reduced necessity of using teeth. One of the main changes in the LN is the development of spindle whorls (Düring 2013). This may show changes in both the techniques of yarn production and the material used for the production of the yarn itself, as well as the produced materials and artefacts such as textiles, rugs and mats, etc. Changes in the frequencies of and increased differences between the sexes in the rates of medial epicondylitis might be the result of changes in the technology of yarn and textile production. Dental grooves have been observed in many prehistoric populations such as the Natufian hunter-foragers (Minozzi et al. 2003; Bocquentin et al. 2005; Eshed et al. 2006). Some examples of dental grooves also come from the Early Neolithic site of Abu Hureyra in the Near East (Molleson 1994). Dental grooves, in general, are common in prehistoric people from the Old and New World (Schulz 1977; Alt, Pichler 1998; Waters-Rist et al. 2010; Lorkiewicz 2011), but activity-induced dental abrasions reduced over time (Waters-Rist et al. 2010; Lorkiewicz 2011). Compared to other published regions, Anatolia has a very high frequency of non-alimentary tooth use, and is one of the best geographical contexts for observing its decreasing frequency together with the development of technology.

Almost all of the previously conducted studies suggest a clear increase in skeletal stress markers from hunter-foragers to agriculturalists. It can even be stated that physical stress in the Late Neolithic communities perhaps reached its highest levels compared to other chronological periods in Anatolia. However, the development of metallurgy, a fully accomplished range of artefacts, long-distance trade and exchange networks and mobility, the formation of centralised administrative systems, the specialisation of production and labour differentiation, the production of surplus, and controlled production and reproduction are characteristic of the LC-EBA period, and resulted in a decreased level of physical stress (Özdoğan 2014; Larsen 2015; Sagona, Zimansky 2015; Frangipane 2017; 2018a; Irvine et al. 2019; Larsen et al. 2019).

Differences between the sexes in workload and the division of labour

One of the most important issues in archaeology is the emergence of differences between the sexes in prehistory. Some studies have identified sex differences in hunter-foragers, albeit at a low level. Moreover, it has been observed that physical stress is a little more evident in females (Eshed et al. 2004; Marchi et al. 2006). It seems that even if physical stresses were related to agrarian lifeways, sex differences in the overall workload were reduced during the transition from hunting and gathering to agriculture (Ruff 1987; Larsen 1995; Panter-Brick 1996; 2002). However, the sexes were affected differently by the absolute increase in physical stress (Ö.D. Erdal 2004; Eshed et al. 2004). For instance, cross-sectional geometric data indicate a marked sex difference in the division of labour with regard to a more symmetrical use of the upper limbs, and different levels of locomotory stress in males and females. For example, more sedentary agricultural tasks have been proposed for western Ligurian females (Marchi et al. 2006).

Eshed et al. (2004) suggested that there were no significant changes in the physical activity of males, while they increased in females with the advent of agriculture. By analysing the higher entheseal scores on the forearms and elbows of Natufian and Neolithic females, it is proposed that women were engaged in activities requiring more dexterity, such as basketry, spinning and weaving. In addition, Molleson (1994) explained the higher values of entheseal changes observed in females at Tell Abu Hureyra (Syria) as being a result of grinding grain in a mortar with a stone or a wooden pestle. Molleson (1989) and Pitre et al. (2017) proposed that severe entheseal changes observed in some arm bones among Neolithic people are associated with grinding activities performed by females. Milella et al. (2012) suggest a possible precocious involvement of females in such types of strenuous activities. A dramatic change was observed not only in western Eurasia but also in the New World. An increase in the bone mass of arm bones in females was interpreted in relation to labour division coinciding with the shift to maize agriculture (Bridges 1983; 1989). The findings of Cassidy (1984) also support this with observed evidence for increased workload in archaic Native American Knoll females. Studies conducted on materials dated to the transition to agriculture in northwestern Alabama (USA) revealed that very few differences existed between male and female foragers compared to western Eurasia. However, male individuals from farming societies displayed a significantly higher level of osteoarthritis than females (Bridges 1991). At Dickson Mound (USA), it seems that with the adoption of agriculture joint diseases increased over time, as well as physical stress, and these were more severe in males (Goodman et al. 1984).

According to the present study, it appears that physical stress increased in both sexes following the adoption of agriculture, albeit in different periods and ways. As suggested above, sex-related differences were slight in the PPNA people, and were characterised by male dominance (Tab. 8.7). However, during the early phase of agriculture, males started to be even more affected, with perimeters, robustness and epicondylitis all showing higher values. This suggests an increased workload for males. Conversely, the osteoarthritis data indicates different conclusions, and females have a high frequency of osteoarthritis in the PPNB. Ö.D. Erdal (2004) suggested that despite the higher frequencies in women, the shoulders, elbows, hips and knees are the most affected joints in men. This observation allows us to suggest sex differences in the division of labour; while the frequency is low in males, together with other data, it shows that the physical load on arms slightly increased from the PPNA to PPNB. A significant change is observed in the hips and knees of women. Degenerative joint diseases in the lower limbs, hips, waists, knees and feet at Abu Hureyra were considered to be the result of squatting to grind cereals (Molleson 1994). The osteoarthritis data and the observed asymmetry for the humerus suggest that with the onset of agriculture females were slightly more affected. However, physical stress related to the forearm is high among males. This may suggest that some physical tasks and activities that produce more physical stress on the forearm – especially movement requiring flexion, pronation and rotation – might be done by males. Making stone bowls, axes and beads, digging up soil and cutting trees using arm power may have been

		Humeral robustness	*Humeral asymmetry*	*Osteoarthritis*	*Entheseal changes*	*Medial epicondylitis*	*Occlusal grooves*
PPNA	Males	Low	Low	Low	Medium	Low	Medium
	Females	Low	Low	Low	Low	Low	High
PPNB	Males	Low	Low	Medium	–	Increased	Low
	Females	Low	High	High	–	Same	Medium
PN	Males	Increased	High	High	–	Decreased	None
	Females	Increased	High	Medium	–	Increased	Medium
LC-EBA	Males	Increased	Low	Medium	–	Decreased	None
	Females	Increased	Low	Low	–	Decreased	Low

Tab. 8.7. Sex differences in selected skeletal characteristics for foragers and agrarian peoples of prehistoric Anatolia.

done by males. Males also contributed to the production of cords and yarns made from plant and animal fibres, as well as textiles, rugs, mats, baskets and other woven objects. The decrease in dental grooves during the PPNB, especially in males, provides important data on the shift from activities requiring the use of the teeth to tasks requiring more physical load on the upper limbs.

The differences in dental abrasion are concomitant with increased asymmetry in the arms, epicondylitis, cupping and arthritis of the shoulder and elbow, and increased sex differences during the PPNB. Osteoarthritis, asymmetry and the robustness of the humeri and radii are more prominent in males during the PN. This suggests that the physical workload on the arms increased significantly in males, and in general in people engaged in intensive agriculture, whereas females dealt with activities requiring more delicate hand movements and dexterity, and developed more medial epicondylitis resulting in physical stresses on the fingers, wrists and elbows. The data suggest that both sexes experienced a slight increase in workload during the early phase of agriculture, while physical pressure on males further increased with the development of intensive farming. Cook (1984) reported, based on data from sites in Illinois in the USA, on similar changes in the patterns of arthritis with the adoption of agriculture.

Another important finding of this study concerns the temporal change in the use of teeth for non-alimentary purposes. There is a wide variety in the abrasions on the dental remains of PPNA people, and different types of abrasion have been previously noted for prehistoric populations (Lorkiewicz 2011). LSAMAT and notches on the anterior teeth are present in every period in Anatolia. However, occlusal grooves exhibit different trends in males and females. Women used their teeth as a tool from the PPNA to the LC-EBA period, with a slight decreasing trend until the LC-EBA period. While men used their teeth intensively in the PPNA, a decrease is observed during the PPNB, and no grooves are observed on the teeth of males after the PPNB. This data may suggest that men moved away from some domestic work requiring manual dexterity, while women continued to work with plant and animal fibres. Molleson (2004) also only found grooves on the teeth of females at Abu Hureyra. Other studies such as Cybulsky (1974) and Y.S. Erdal (2008) suggested that only women used their anterior teeth to process fibres from plants and animals. Waters-Rist (2010) found that grooves are found predominantly in females in Holocene Siberian populations, and Lorkiewicz (2011) found them predominantly in females of the Lengyel Culture in Poland.

Occlusal grooves likely resulted from yarn and cord production in Anatolia (Y.S. Erdal 2008). Weaving, spinning and grinding are highly repetitive tasks, as demonstrated by ethnographic and anthropological studies on contemporary people, and these tasks were most probably practised by women in the past (Y.S. Erdal 2008; Üstündağ 2020). Cook (1984) suggested that women may have experienced an intensification of their activity patterns, and whilst new duties might have been added to their usual daily activities, the findings of this study also indicate a continuity in the work patterns of females. Conversely, increased physical stress on male bones suggests that men could have experienced a more physically stressful change in their activities. These observations imply increased differences between the sexes with the onset of agriculture in Anatolia.

Conclusion

Archaeological and physical anthropology studies have proposed that the PPN period was characterised by a transition from a hunter-forager lifestyle to an agricultural one. The secondary effects of the Neolithisation process may be listed as follows: an increase in the number and size of settlements, population growth, changes in the quality of foods such as a decrease in animal protein intake, an increase in carbohydrate consumption and a narrowing in the range of exploited food resources. Studies on demographic patterns also suggest changes in morbidity and mortality. All of these transformations were related to the transition to new subsistence strategies, dietary habits, social organisation and rituals (Bar-Yosef, Belfer-Cohen 1989; Banning 1998; Bar-Yosef 2001). All of these changes that occurred in the daily life of Neolithic populations likely resulted in increased physical stresses and workloads. Ethnographic data from Anatolian villages show that both men and women are active in agricultural activities and other works, albeit using different tools (Üstündağ 2020). With the adoption of agriculture, both sexes underwent greater physical stresses, but in different ways, by using different tools and by performing different particular and distinct tasks and activities. As suggested by Eshed et al. (2004), the division of labour between the sexes became more evident during the transition to agriculture. Agrarian Neolithic people were engaged in a wider variety of occupational activities compared to the hunter-foragers of western Eurasia. Females were more affected at the onset of agriculture, while males were the most affected following the full adoption and intensification of domestic agriculture. Furthermore, it is possible to claim that sex differences in labour and habitual tasks and activities was one of the main effects of the Neolithisation process on Anatolian populations, following examination of their skeletal remains.

Acknowledgements

This study is based on data collected by researchers and colleagues who have worked with me in the Hacettepe University Skeletal Biology Laboratory on different skeletal populations and societies from the Neolithic to Early Bronze Age periods. I would like to thank the archaeologists who have worked with me for a long time, and my students who have provided information from their theses. However, I am particularly grateful to Valentina D'Amico, Demet Delibaş and Meliha Melis Koruyucu, who helped to improve this work with their reading and editing of the manuscript. I would also like to thank Dr Nefize Ezgi Altınışık, who prepared the high-resolution map and graphs for this manuscript.

Bibliography

Algaze, G., Matney, T. 2011: 'Titriş Höyük' in S.R. Steadman, G. McMahon (eds), *Oxford Handbook of Anatolian Studies*. Oxford, Oxford University Press: 994–1011

al-Oumaoui I., Brobeil, J.S., du Souich, P. 2004: 'Markers of activity patterns in some populations of the Iberian Peninsula' *International Journal of Osteoarchaeology* 14: 343–59

Alt, K.W., Pichler, S.L. 1998: 'Artificial modifications of human teeth' in K.W. Alt, F.W. Rösing, M. Teschler-Nicola (eds), *Dental Anthropology: Fundamentals, Limits, and Prospects*. Vienna, Springer: 387–415

Angel, J.L. 1984: 'Health as a crucial factor in the changes from hunting to developed farming in the eastern Mediterranean' in M.N. Cohen, G.J. Armelagos (eds), *Paleopathology at the Origins of Agriculture*. Orlando, Academic Press: 51–74

Arbuckle, B.S., Özkaya, V. 2006: 'Animal exploitation at Körtik Tepe: an Early Aceramic Neolithic site in southeastern Turkey' *Paléorient* 32.2: 113–36

Arensburg, B. 1985: 'A short review of paleopathology in the Middle East. Mitekufat Haeven' *Journal of the Israel Prehistoric Society* 18: 21–36

Armelagos, G.J., Cohen, M.N., eds. 1984: *Paleopathology at the Origins of Agriculture*. Orlando, Academic Press

Aslan, F.G. 2005: *Anadolu'da neolitik dönemden günümüze kalan uzun kemiklerin boyutları ve iskelet kütleviliğinin değerlendirilmesi*. PhD thesis, Hacettepe University, Ankara

Auerbach, B.M., Ruff, C.B. 2006: 'Limb bone bilateral asymmetry: variability and commonality among modern humans' *Journal of Human Evolution* 50: 203–18

Bamyacı, O. 2018. 'Öğütme Taşları: tipolojik gelişim, sorunlar, çalışma metodolojisi ve analitik bir yaklaşım modeli' in A. Baysal (ed.) *Anadolu Arkeolojisinde Taş Aletler: Teori, Metot, Pratik*. İstanbul, Ege Yayınları: 271–302

Banning, E.B. 1998: 'The Neolithic period: triumphs of architecture, agriculture, and art' *Near Eastern Archaeology* 61.4: 188–237

Bar-Yosef, O. 2001: 'From sedentary foragers to village hierarchies: the emergence of social institutions' *Proceedings of the British Academy* 110: 1–38

Bar-Yosef, O., Belfer-Cohen, A. 1989: 'The origins of sedentism and farming communities in the Levant' *Journal of World Prehistory* 3.4: 447–98

Belfer-Cohen, A., Bar-Yosef, O. 2000: 'Early sedentism in the Near East: a bumpy ride to village life' in I. Kuijt (ed.), *Life in Neolithic Farming Communities: Social Organization, Identity and Differentiation*. New York, Kluwer Academic/Plenum Publishers: 19–37

Benz, M., Deckers, K., Rössner, C., Alexandrovskiy, A., Pustovoytov, K., Scheeres, M., Fecher, M., Coşkun, A., Riehl, S., Alt, K.W., Özkaya, V. 2015: 'Prelude to village life. Environmental data and building traditions of the Epipalaeolithic settlement at Körtik Tepe, southeastern Turkey' *Paléorient* 41.2: 9–30

Benz, M., Erdal, Y.S., Şahin, F.S., Özkaya, V., Alt, K.W. 2016a: 'The equality of inequality. Social differentiation among the foragers of Körtik Tepe, southeastern Turkey' in H. Mellart, H.P. Hahn, R. Jung, R. Risch (eds), *Rich and Poor – Competing for Resources in Prehistory. Tagungen des Landesmuseums für Vorgeschichte Halle 13*. Halle: Landesamt für Denkmalpflege und Archäologie Sachsen Anhalt, Landesmuseum für Vorgeschichte Halle (Saale): 147–64

Benz, M., Fecher, M., Scheeres, M., Alt, K.W., Erdal, Y.S., Şahin, F.S., Özkaya, V. 2016b: 'Results of stable isotopes from Körtik Tepe southeastern Turkey' *Arkeometri Sonuçları Toplantısı* 31: 231–52

Bernbeck, R., Nieuwenhuyse, O.P. 2013: 'Established paradigms, current disputes and emerging themes: the state of research on the Late Neolithic in Upper Mesopotamia' in O.P. Nieuwenhuyse, R. Bernbeck, P.M.M.G. Akkermans, J. Rogasch (eds), *Interpreting the Late Neolithic of Upper Mesopotamia*. Tournai, Brepols Publishers: 17–37

Bilgi, Ö. 1984: 'Metal objects from İkiztepe-Turkey' *Beiträge zur Allgemeinen und Vergleichenden Archäologie* 6: 31–97

— 1990 'Metal objects from İkiztepe-Turkey' *Beiträge zur Allgemeinen und Vergleichenden Archäologie* 9–10: 119–219

Bocquentin, F., Sellier, P., Murail, P. 2005: 'Abrasion dentaire et travail spécialisé dans la population natoufienne de Mallaha (Israël)' *Comptes Rendus Palevol* 4.4: 351–57

Bogin, B. 2001: *The Growth of Humanity*. New York: Wiley Liss

Bridges, P.S. 1983: 'Subsistence activities and biomechanical properties of long bones in 2 Amerindian populations' *American Journal of Physical Anthropology* 60.2: 177

Bridges, P.S. 1989: 'Changes in activities with the shift to agriculture in the southeastern United States' *Current Anthropology* 30.3: 385–94

— 1991: 'Degenerative joint disease in hunter-gatherers and agriculturalists from the southeastern United States' *American Journal of Physical Anthropology* 85.4: 379–91

Buikstra, J.E., Ubelaker, D. 1994: *Standards for Data Collection from Human Skeletal Remains*. Fayetteville, Arkansas Archaeological Survey

Buitenhuis, H., Peters, J., Pöllath, N., Stiner, M.C., Munro, N.D., Saritaş, Ö. 2018: 'The faunal remains from levels 3 and 2 of Aşıklı Hoyuk: evidence for emerging management practices' in M. Özbaşaran, G. Duru, M. Stiner (eds) *The Early Settlement of Aşıklı Höyük: Essays in Honor of Ufuk Esin*. Ege Yayınları, İstanbul: 282–323

Byrd, B.F. 2005: 'Reassessing the emergence of village life in the Near East' *Journal of Archaeological Research* 13.3: 231–90

Cassidy, C.M. 1984: 'Skeletal evidence for prehistoric subsistence adaptation in the Central Ohio River Valley' in M.N. Cohen, G.J. Armelagos (eds) *Paleopathology at the Origins of Agriculture*. Orlando, Academic Press: 307–46

Çevik, Ö. 2007: 'The emergence of different social systems in Early Bronze Age Anatolia: urbanisation versus centralisation' *Anatolian Studies* 57: 131–40

Çırak, A., Çırak, M.T., Erkman, A.C. 2013: 'Paleopathological analysis of teeth and jaws obtained from Kelenderis excavations' *OLBA* 21: 1–23

Cook, D.C. 1984: 'Subsistence and health in the Lower Illinois Valley: osteological evidence' in M.N. Cohen, G.J. Armelagos (eds) *Paleopathology at the Origins of Agriculture*. Orlando, Academic Press: 235–69

Coşkun A., Benz, M., Erdal, Y.S., Koruyucu, M.M., Deckers, K., Riehl, S., Siebert, A., Alt, K.W., Özkaya, V. 2011: 'Living by the water – boon and bane for the people of Körtik Tepe' *NeoLithics* 2.10: 60–71

Crane, A., Watson, J.T., Haas, R. 2020: 'The effects of aberrant tooth wear on occlusal relationships' in C.W. Schmidt, J.T. Watson (eds), *Dental Wear in Evolutionary and Biocultural Contexts*. London, Academic Press: 99–121

Crianni, R., Fornaciari, G. 2003: 'Luigi Boccherini and the Barocco Cello: an 18th century striking case of occupational disease' *International Journal of Osteoarchaeology* 13: 294–302

Delibaş, D. 2016: *Körtik Tepe insanlarında gündelik yaşam: İskelet kas tutunma bölgelerinin analizleriyle hareket biçiminin belirlenmesi*. MA thesis, Hacettepe University, Ankara.

Duru, G. 2015: 'Kamusal insanın (ilk) çöküşü' in A. Baysal (ed.). *İletişim Ağları ve Sosyal Organizasyon. Tematik Arkeoloji Serisi 2*. İstanbul: Ege Yayınları: 195–210

Duru, R., Umurtak, G. 2012: 'Bademağacı kazıları 2010 yılı çalışma raporu'. *3. Kazı Sonuçları Toplantısı* 2: 115–24

Dutour, O. 1986: 'Enthesopathies (lesions of muscular insertions) as indicators of the activities of Neolithic Saharan populations' *American Journal of Physical Anthropology* 71.2: 221–24

Duyar, İ., Özener, B. 2003: *Çocuk İşçiler: Çarpık Gelişen Bedenler*. Ankara, Ütopya Yayınevi

Düring, B.S. 2013. 'Apples or oranges? Comparing Neolithic horizons of Upper Mesopotamia and Asia Minor' in O.P. Nieuwenhuyse, R. Bernbeck, P.M.M.G. Akkermans, J. Rogasch (eds), *Interpreting the Late Neolithic of Upper Mesopotamia*. Tournai, Brepols Publishers: 367–76

Düring, B.S., Marciniak, A. 2005: 'Households and communities in the central Anatolian Neolithic' *Archaeological Dialogues* 12: 165–87

Efe, T., Fidan, M.E. 2006: 'Pre-Middle Bronze Age metal objects from inland Western Anatolia: a typological and chronological evaluation' *Anatolia Antiqua. Eski Anadolu*, 14.1: 15–43

Erdal, Ö.D. 2004: *Eklem hastalıklarının yaşam biçimiyle ilişkisi: Eski Anadolu toplulukları örneği*. PhD thesis, Hacettepe University, Ankara

Erdal, Y.S. 1999: 'Neolitikten günümüze Anadolu insan topluluklarının dişlerinde boyut küçülmesi' *Hacettepe Diş Hekimliği Fakültesi Dergisi* 23.1: 12–29

— 2008: 'Occlusal grooves in anterior dentition among Kovuklukaya inhabitants (Sinop, Northern Anatolia, 10th century AD)' *International Journal of Osteoarchaeology* 18.2: 152–66

— 2013: 'Life and death at Hakemi Use' in O.P. Nieuwenhuyse, R. Bernbeck, P.M.M.G. Akkermans, J. Rogasch (eds), *Interpreting the Late Neolithic of Upper Mesopotamia*. Tournai, Brepols Publishers: 213–23

— 2015: 'Bone or flesh: defleshing and post-depositional treatments at Körtik Tepe (southeastern Anatolia, PPNA Period)' *European Journal of Archaeology* 18.1: 4–32

— 2019b: 'Middle Chalcolithic burials from Yumuktepe' in G. Palumbi, I. Caneva (eds), *The Chalcolithic at Mersin-Yumuktepe Level XVI Reconsidered*. Istanbul, Ege Yayınları: 181–83

— 2021: 'Tooth as a tool: activity induced dental abrasion in prehistoric Anatolia' Paleorient 47.1: 125–49

Erdal Y.S., Takaoğlu, T. 2021: 'Mortuary behavior in Chalcolithic Anatolia: a view from Gülpınar' *Turkish Journal of Archaeological Sciences* 1: 42–66

Erim-Özdoğan, A. 2011: 'Çayönü' in M. Özdoğan, N. Başgelen, P. Kuniholm (eds), *The Neolithic in Turkey – The Tigris Basin: New Excavations & New Research*. Istanbul, İstanbul Arkeoloji ve Sanat Yayınları: 185–269

Eshed, V., Gopher, A., Gage, T.B., Hershkovitz, I. 2003: 'Has the transition to agriculture reshaped the demographic structure of prehistoric populations? New evidence from the Levant' *American Journal of Physical Anthropology* 124: 315–29

Eshed, V., Gopher, A., Galili, E., Hershkovitz, I. 2004: 'Musculoskeletal stress markers in Natufian hunter-gatherers and Neolithic farmers in the Levant: the upper limb' *American Journal of Physical Anthropology* 123: 303–15

Eshed, V., Gopher, A., Hershkovitz, I. 2006: 'Tooth wear and dental pathology at the advent of agriculture: new evidence from the Levant' *American Journal of Physical Anthropology* 130: 145–59

Eshed, V., Gopher, A., Pinhasi, R., Hershkovitz, I. 2010: 'Paleopathology and the origin of agriculture in the Levant' *American Journal of Physical Anthropology* 143.1: 121–33

Evershed, R.P., Payne, S., Sherratt, A.G. Copley, M.S., Coolidge, J., Urem-Kotsu, F., Kotsakis, K., Ozdoğan, M., Ozdoğan, S.A., Nieuwenhuyse, O., Akkermans, P.M.M.G., Bailey, D., Andeescu, R.R., Campbell, S., Farid, S., Hodder, I., Yalman, N., Ozbaşaran, M., Bıçakcı, E., Garfinkel, Y., Levy, T. Burton, M.M. 2008: 'Earliest date of milk use in the Near East and Southern Europe linked to cattle herding' *Nature* 55: 528–31

Frangipane, M. 2002: *Yakındoğu'da Devletin Doğuşu*. Istanbul, Arkeoloji ve Sanat Yayınları

— 2017: 'The role of metal procurement in the wide interregional connections of Arslantepe during the late 4th–early 3rd millennia BC' in Ç. Maner, M.T. Horowitz, A.S. Gilbert (eds) *Overturning Certainties in Near Eastern Archaeology*. Leiden, Brill: 186–210

Frayer, D.W., Wolpoff, M.H. 1985: 'Sexual dimorphism' *Annual Review of Anthropology*, 14.1: 429–73

Goodman, A.H., Lallo. J., Armelagos, G.J., Rose, J.C. 1984: 'Health change at Dickson Mounds, Illinois (A.D. 950–1300)' in M.N. Cohen, G.J. Armelagos (eds), *Paleopathology at the Origins of Agriculture*. Orlando, Academic Press: 271–306

Güleç, E. 1989: 'Panaztepe iskeletlerinin paleoantropolojik ve paleopatolojik incelemesi' *Türk Arkeoloji Dergisi* 28: 73–95

Gündoğan, Ü., Şahoğlu, V., Erkanal, H. 2019: 'Spatial organization and production activities at Bakla Tepe during the first half of the 3rd millennium BC' *Dil ve Tarih-Coğrafya Fakültesi Dergisi* 59.2: 1087–110

Hawkey, D.E. Merbs, C.F. 1995: 'Activity-induced Musculoskeletal Stress Markers (MSM) and subsistence strategy changes among ancient Hudson Bay Eskimos' *International Journal of Osteoarchaeology* 5: 324–38

Irvine, B., Thomas, J.L. Schoop, U.-D. 2014: 'A macroscopic analysis of human dentition at Late Chalcolithic Çamlıbel Tarlası, North Central Anatolia, with special reference to dietary and non-masticatory habits' *Interdisciplinaria Archaeologica* 5.1: 19–30

Irvine, B., Erdal, Y.S., Richards, M.P. 2019: 'Dietary habits in the Early Bronze Age (3rd millennium BC) of Anatolia: a multi-isotopic approach' *Journal of Archaeological Science*: Reports 24: 253–63

Itahashi, Y., Erdal, Y.S., Tekin, H., Omar, L., Miyake, Y., Chikaraishi, Y., Ohkouchi, N., Yoneda, M. 2019: 'Amino acid 15N analysis reveals change in the importance of freshwater resources between the hunter-gatherer and farmer in the Neolithic upper Tigris' *American Journal of Physical Anthropology* 168.4: 676–86

Jones, H.H., Priest, J.D., Hayes, W.C., Tichenor, C.C., Nagel, D.A. 1977: 'Humeral hypertrophy in response to exercise' *The Journal of Bone and Joint Surgery American* 59.2: 204–08

Jurmain, R.D. 1977a: 'Stress and the etiology of osteoarthritis' *American Journal of Physical Anthropology* 46.2: 353–65

— 1977b: 'Part one: paleoepidemiology of degenerative knee disease' *Medical Anthropology* 1.1: 1–23

— 1990: 'Paleoepidemiology of a central California prehistoric population from CA-ALA-329: II. Degenerative disease' *American Journal of Physical Anthropology* 83.1: 83–94

Kennedy, K.A. 1983: 'Morphological variations in ulnar supinator crests and fossae as identifying markers of occupational stress' *Journal of Forensic Science* 28.4: 871–76

— 1984: 'Biological adaptations and affinities of Mesolithic South Asians' in J.R. Lukacs (ed.), *The People of South Asia*. Boston, Springer: 29–57

— 1989: 'Skeletal markers of occupational stress' in M.Y. Iscan, K.A.R. Kennedy (eds), *Reconstruction of Life from the Skeleton*. New York, Alan R. Liss: 129–61

Knüsel, C. 2000: 'Bone adaptation and its relationship to physical activity in the past' *Human Osteology in Archaeology and Forensic Science*: 381–401

Koca Özer, B., Güleç, E., Gültekin, T., Akın, G., Güngör, K., Brooks, S.L. 2006: 'Implications of dental caries in Anatolia from hunting gathering to the present' *Human Evolution* 21: 215–22

Kodaş, E., Sağlamtimur, H., Erdal, Y.S. 2018: 'Three human graves of the Hassuna Culture in Türbe Höyük' *Anatolia Antiqua. Revue internationale d'archéologie anatolienne* 26: 13–21

Koruyucu, M.M., Şahin, F.S., Delibaş, D., Erdal, Ö.D., Benz, M., Özkaya, V., Erdal, Y.S. 2018: 'Auditory exostosis: exploring daily life at an early sedentary population (Körtik Tepe, Turkey)' *International Journal of Osteoarchaeology* 28.6: 615–25

Laneri, N. 2007: 'Burial practices at Titriş Höyük, Turkey: an interpretation' *Journal of Near Eastern Studies* 66: 241–66

Larsen, C.S. 1995: 'Biological changes in human populations with agriculture' *Annual Review of Anthropology* 24.1: 185–213

— 1997: 'Bare bones anthropology: the bioarchaeology of human remains' *Research Frontiers in Anthropology*: 3–24

— 2015: *Bioarchaeology: Interpreting Behavior from the Human Skeleton*. Cambridge, Cambridge University Press

Larsen, C.S., Hillson, S.W., Boz, B., Pilloud, M.A., Sadvari, J.W., Agarwal, S.C., Glencross, B., Beauchesne, P., Pearson, J., Rıff, C.B., Garofalo, E.M., Hager, L.D., Haddow, S.D., Knüsel, C.J. 2015: 'Bioarchaeology of Neolithic Çatalhöyük: lives and lifestyles of an early farming society in transition' *Journal of World Prehistory* 28.1: 27–68

Larsen, C.S., Knüsel, C.J., Haddow, S.D., Pilloud, M.A., Milella, M., Sadvari, J.W., Pearson, J., Ruff, C.B., Garofalo, E.M., Bocaehe, E., Betz, B.J, Dori, I., Glencross, B. 2019: 'Bioarchaeology of Neolithic Çatalhöyük reveals fundamental transitions in health, mobility, and lifestyle in early farmers' *Proceedings of the National Academy of Sciences* 116.26: 12615–23

Lorkiewicz, W. 2011: 'Non alimentary tooth use in the Neolithic population of the Lengyel Culture in Central Poland (4600–4000 BC)' *American Journal of Physical Anthropology* 144.4: 538–51

Lovell, N.C., Dublenko, A.A. 1999: 'Further aspects of fur trade life depicted in the skeleton' *International Journal of Osteoarchaeology* 9.4: 248–56

Mahoney, E.K., Kilpatrick, N.M. 2003: 'Dental erosion: part 1. Aetiology and prevalence of dental erosion' *The New Zealand Dental Journal* 99.2: 33–41

Marchi, D., Sparacello, V.S., Holt, B.M., Formicola, V. 2006: 'Biomechanical approach to the reconstruction of activity patterns in Neolithic Western Liguria, Italy' *American Journal of Physical Anthropology* 131: 447–55

Mark, Y., Kate, H., Adam, T., Srinath, K. 2015: 'Lateral and medial epicondylitis' *MedCrave Online Journal of Orthopedics & Rheumatology* 3.2: 00090

Marklin, R.W., Monroe, J.F. 1998: 'Quantitative biomechanical analysis of wrist motion in bone trimming jobs in the meatpacking industry' *Ergonomics* 41: 227–37

Matney, T., Algaze, G. 1995: 'Urban development at Mid-Late Early Bronze Age Titriş Höyük in southeastern Anatolia' *Bulletin of the American Schools of Oriental Research* 299/300: 33–52

Merbs, C.F. 1983: *Patterns of Activity-induced Pathology in a Canadian Inuit Population*. Ottawa, National Museum of Canada

Milella, M., Giovanna Belcastro, M., Zollikofer, C.P., Mariotti, V. 2012: 'The effect of age, sex, and physical activity on entheseal morphology in a contemporary Italian skeletal collection' *American Journal of Physical Anthropology* 148.3: 379–88

Miller, R.J. 1985: 'Lateral epicondylitis in the prehistoric Indian population from Nuvakwewtaqa (Chavez Pass), Arizona' *Health and Diseases in the Prehistoric Southwest* 34: 391–400

Minozzi, S., Manzi, G., Ricci, F., di Lernia, S., Borgognini Tarli, S.M. 2003: 'Non alimentary tooth use in prehistory: an example from early Holocene in Central Sahara (Uan Muhuggiag, Tadrart Acacus, Libya)' *American Journal of Physical Anthropology* 120.3: 225–32

Molleson, T. 1989: 'Seed preparation in the Mesolithic: the osteological evidence' *Antiquity*, 63.239: 356–62
— 1994: 'The eloquent bones of Abu Hureyra' *Scientific American* 271.2: 60–65
— 2000: 'The people of Abu Hureyra' in A.M.T. Moore, G.C. Hillman, A.J. Legge (eds) *Village on the Euphrates: From Foraging to Farming at Abu Hureyra*. Oxford, Oxford University Press: 301–24
Molleson, T., Blondiaux, J. 1994: 'Riders' bones from Kish, Iraq' *Cambridge Archaeological Journal* 4.2: 312–16
Molleson, T., Hodgson, D. 2000: 'The porters of Ur' *Isimu* 3: 101–17
Molnar, S. 1971: 'Human tooth wear, tooth function and cultural variability' *American Journal of Physical Anthropology* 34.2: 175–89
Molnar, P. 2010: 'Patterns of physical activity and material culture in Gotland, Sweden, during the Middle Neolithic' *International Journal of Osteoarchaeology* 20: 1–14
Mummert, A., Esche, E., Robinson, J., Armelagos, G. J. 2011: 'Stature and robusticity during the agricultural transition: evidence from the bioarchaeological record' *Economics & Human Biology* 9.3: 284–301
Nilsson, B.E., Westlin, N.E. 1971: 'Bone density in athletes' *Clinical Orthopaedics and Related Research* (1976–2007) 77: 179–82
Ortner, D.J. 2003: *Identification of Pathological Conditions in Human Skeletal Remains*. Amsterdam, Academic Press
Öz, E. 2012: 'Kültepe tabletlerinde geçen tarım aletleri Saban (epinnum) ve Orak (niggallum)' *Arkeoloji ve Sanat* 140: 21–24
Özbaşaran, M., Duru, G., Stiner, M.C., Esin, U. 2018: *The Early Settlement at Aşıklı Höyük: Essays in Honor of Ufuk Esin*. Istanbul, Ege Yayınları
Özbek, M. 2004: *Çayönü'nde İnsan*. İstanbul, Arkeoloji ve Sanat Yayınları
— 2012: 'Auditory exostoses among the Prepottery Neolithic inhabitants of Çayönü and Aşıklı, Anatolia, its relation to aquatic activities' *International Journal of Paleopathology* 2.4: 181–86
Özdogan, M. 2014: 'Anatolia: from the pre-pottery Neolithic to the end of the Early Bronze Age (10,500–2000 BCE)' in C. Renfrew, P. Bahn (eds), *The Cambridge World Prehistory*. Cambridge, Cambridge University Press: 1508–44
Özdoğan, M., Özdoğan, A. 1989: 'Çayönü: a conspectus of recent work' *Paléorient*: 65–74
Özener, B. 2007: 'Biomechanical pressures and upper extremity asymmetry: a study on young laborers' *Collegium Antropologicum* 31.3: 693–99
— 2010: 'Fluctuating and directional asymmetry in young human males: effect of heavy working condition and socioeconomic status' *American Journal of Physical Anthropology* 143.1: 112–20
Özer, B.K., Sağır, M., Özer, İ. 2011: 'Secular changes in the height of the inhabitants of Anatolia (Turkey) from the 10th millennium BC to the 20th century AD' *Economics & Human Biology* 9.2: 211–19
Özkaya, V. 2009: 'Excavations at Körtik Tepe. A new Pre-Pottery Neolithic site in southeastern Anatolia' *NeoLithics* 2: 3–8
Özkaya, V., Coşkun, A. 2007: 'Körtik Tepe' in N. Başgelen (ed.), *12000 Yıl Önce 'Uygarlığın Anadolu'dan Avrupa'ya Yolculuğunun Başlangıcı' Neolitik Dönem*. İstanbul, Yapı Kredi Yayınları: 86–88
Özkaya, V., Coşkun, A. 2011: 'Körtik Tepe' in: M. Özdoğan, N. Başgelen, P. Kuniholm (eds), *The Neolithic in Turkey – The Tigris Basin: New Excavations & New Research*. Istanbul, Arkeoloji ve Sanat Yayınları: 89–127
Özkaya, V., Coşkun, A., Benz, M., Erdal, Y.S., Atıcı, L. 2012: 'Körtik Tepe 2010 Kazısı' *Kazı Sonuçları Toplantısı* 33: 315–38
Özkaya, V., Coşkun, A., San, O., Şahin, F.S., Barın, G., Kartal, M., Erdal, Y.S. 2010: 'Körtik Tepe 2008 Kazısı' *Kazı Sonuçları Toplantısı* 31.1: 511–35
Panter-Brick, C. 1996: 'Seasonal and sex variation in physical activity levels among agro-pastoralists in Nepal' *American Journal of Physical Anthropology* 100: 7–21
— 2002: 'Sexual division of labor: energetic and evolutionary scenarios' *American Journal of Human Biology* 14: 627–40
Pascarelli, E.F., Hsu, Y.P. 2001: 'Understanding work-related upper extremity disorders: clinical findings in 485 computer users, musicians, and others' *Journal of Occupational Rehabilitation* 11: 1–21
Peterson, J. 1997: 'Tracking activity patterns through skeletal remains: a case study from Jordan and Palestine' in H.G.K. Gebel, Z. Kafafi, G.O. Rollefson (eds), *The Prehistory of Jordan II. Perspectives from 1997. Studies in Early Near Eastern Production, Subsistence, and Environment* 4. Berlin, Ex Orient: 475–92
Pitre, M.C., Koliński, R., Sołtysiak, A. 2017: 'A possible "grinder" from Tell Arbid, Syria' *Anthropologischer Anzeiger; Bericht uber die biologisch-anthropologische Literatur* 74.4: 297–307

Priest, J.D., Jones, H.H., Tichenor, C.J.C., Nagel, D.A. 1977: 'Arm and elbow changes in expert tennis players' *Minnesota Medicine* 60.5: 399–404

Rathbun, T.A. 1984: 'Skeletal pathology from the Palaeolithic through the Metal Ages in Iran and Iraq' in M. Cohen, G.J. Armelagos (eds), *Paleopathology at the Origins of Agriculture*. Orlando, Academic Press: 137–68

Riehl, S., Benz, M., Conard, N.J., Darabi, H., Deckers, K., Nashli, H.F., Zeidi-Kulehparcheh, M. 2011: 'Plant use in three Pre-Pottery Neolithic sites of the northern and eastern Fertile Crescent: a preliminary report' *Vegetation History and Archaeobotany* 21: 95–106

Rosenstock, E., Ebert, J., Martin, R., Hickeiter, A. Walter, P., Groß, M. 2019: 'Human stature in the Near East and Europe ca. 10,000–1000 BC: its spatiotemporal development in a Bayesian errors-in-variables model' *Archaeological and Anthropological Sciences* 11: 5657–90

Rothman, M.S. 2017: 'Explaining the Kura-Araxes' in A.T. Smith, K.O. Weber, E. Hite, L. Khatchadourian (eds) *Fitful Histories and Unruly Publics: Rethinking Temporality and Community in Eurasian Archaeology*. Boston, Brill: 217–57

Ruff, C. 1987: 'Sexual dimorphism in human lower limb bone structure: relationship to subsistence strategy and sexual division of labor' *Journal of Human Evolution* 16.5: 391–416

— 2005: 'Mechanical determinants of bone form: insights from skeletal remains' *Journal of Musculoskeletal Neuronal Interactions* 5.3: 202–12

Ruff, C.B., Jones, H.H. 1981: 'Bilateral asymmetry in cortical bone of the humerus and tibia-sex and age factors' *Human Biology*: 69–86

Ruff, C.B., Larsen, C.S., Hayes, W.C. 1984: 'Structural changes in the femur with the transition to agriculture on the Georgia Coast' *American Journal of Physical Anthropology* 64.2: 125–36

Santana, J., Velasco, J., Balbo, A., Iriarte, E., Zapata, L., Teira, L., Nicolle, C., Braemer, F., Ibáñez, J.J. 2015: 'Interpreting a ritual funerary area at the Early Neolithic site of Tell Qarassa North (South Syria, late 9th millennium BC)' *Journal of Anthropological Archaeology* 37: 112–27

Schulz, P.D. 1977: 'Task activity and anterior tooth grooving in prehistoric California Indians' *American Journal of Physical Anthropology* 46.1: 87–91

Shuler, K.A., Zeng, P., Danforth, M.E. 2012: 'Upper limb entheseal change with the transition to agriculture in the southeastern United States: a view from Moundville and the Central Tombigbee River Valley' *HOMO – Journal of Comparative Human Biology* 63.6: 413–34

Skourtanioti, E., Erdal, Y.S., Frangipane, M., Restelli, F.B., Yener, K.A., Pinnock, Matthiae, P., Özbal, R, Schoop, U.-D., Guliyev, F, Akhundov, T., Lyonet, B., Hammer, E.L., Nugent, S.E., Burri, M., Gunnar, U.N., Penske, S., Ingman, T., Akar, M., Shafiq, R., Palumbi, G., Einsenman ,S., D'Andrea, M., Rohrlach, A., Warrinner, C., Choongwon, J., Stockhammer, P.W., Haak, W., Krause, J. 2020: 'Genomic history of Neolithic to Bronze Age Anatolia, Northern Levant, and Southern Caucasus' *Cell* 181.5: 1158–75

Smith, P., Horwitz, L.K. 2007: 'Ancestors and inheritors: a bioarchaeological perspective on the transition to agropastoralism in the southern Levant' in M.N. Cohen, G.M.M. Crane-Kramer (eds), *Ancient Health: Skeletal Indicators of Agricultural and Economic Intensification*. Gainesville, University Press of Florida: 207–22

Spigelman, M., Erdal, Y.S., Donoghue, H.D. Pinhasi, R. 2012: 'Golfer and tennis elbow in Byzantine Turkey: epicondylitis a neglected occupation/activity marker in antiquity' *Advances in Anthropology* 2.1: 24–30

Susanne, C. 1977: 'Heritability of anthropological characters' *Human Biology*: 573–80

Steen, L.S., Lane, R.W. 1998: 'Evaluation of habitual activities among two Alaskan Eskimo populations based on musculoskeletal stress markers' *International Journal of Osteoarchaeology* 8: 341–53

Steinbock, R.T. 1976: *Paleopathological Diagnosis and Interpretation: Bone Diseases in Ancient Human Populations*. Springfield, IL, Charles C. Thomas Publisher Limited

Şahoğlu, V. 2005: 'The Anatolian trade network and the Izmir region during the Early Bronze Age' *Oxford Journal of Archaeology* 24.4: 339–61

Shiri, R., Viikari-Juntura, E., Varonen, H., Heliövaara, M. 2006: 'Prevalence and determinants of lateral and medial epicondylitis: a population study' *American Journal of Epidemiology* 164.11:1065–74

Tekin, H. 2011: 'Hakemi Use, a new discovered Late Neolithic site in southeastern Anatolia' in M. Özdoğan, N. Başgelen, P. Kuniholm (eds), *The Neolithic in Turkey – The Tigris Basin: New Excavations & New Research*. Istanbul, Arkeoloji ve Sanat Yayınları: 151–72

— 2013: 'The contribution of Hakemi Use to Upper Mesopotamia' in O.P. Niuewenhuyse, R. Bernbeck, P.M.M.G. Akkermans, J. Rogasch (eds), *Interpreting the Late Neolithic of Upper Mesopotamia*. Tournai, Brepols Publishers: 493–502

Tomczyk, J., Zalewska, M. 2016: 'Mechanical and chemical dental wear in the historical population from the Syrian lower Euphrates valley' *Archives of Oral Biology* 62: 49–57

Trinkaus, E., Churchill, S.E., Ruff, C.B. 1994: 'Postcranial robusticity in Homo. II: humeral bilateral asymmetry and bone plasticity' *American Journal of Physical Anthropology* 93.1: 1–34

Tschantz, P., Meine, J. 1993: 'Medial epicondylitis. Etiology, diagnosis, therapeutic modalities' *Zeitschrift fur Unfall-chirurgie und Versicherungsmedizin: Offizielles Organ der Schweizerischen Gesellschaft fur Unfallmedizin und Berufskrankheiten Revue de Traumatologie et D'assicurologie: Organe Officiel de la Societe Suisse* 86.3: 145–48

Üstündağ, H. 2020: 'Entheseal changes in the Hellenistic-Roman population of Boğazköy, Turkey: evidence for gender division of labor?' *Bioarchaeology of the Near East* 14: 1–23

Waldron, T. 1997: 'Osteoarthritis of the hip in past populations' *International Journal of Osteoarchaeology* 7.2: 186–89

Walker-Bone, K., Palmer, K.T., Reading, I., Coggon, D., Copper, C. 2004: 'Prevalence and impact of musculoskeletal disorders of the upper limb in the general population' *Arthritis and Rheumatism* 51: 642–51

Walz, D.M., Newman, J.S., Konin, G.P., Ross, G. 2010: 'Epicondylitis: pathogenesis, imaging, and treatment' *Radio-graphics* 30.1: 167–84

Waters-Rist, A., Bazaliiskii, V.I., Weber, A., Goriunova, O.I., Katzenberg, M.A. 2010: 'Activity-induced dental modification in Holocene Siberian hunter-fisher-gatherers' *American Journal of Physical Anthropology* 143.2: 266–78

Watson, J.T., Haas, R. 2017: 'Dental evidence for wild tuber processing among Titicaca Basin foragers 7000 YBP' *American Journal of Physical Anthropology* 164.1: 117–30

Weiss, E. 2007: 'Muscle markers revisited: activity pattern reconstruction with controls in a Central California Amerind population' *American Journal of Physical Anthropology* 133: 931–40

— 2009: *Bioarchaeological Science: What We Have Learned from Human Skeletal Remains*. New York, Nova Science Publisher

Welton, M.L. 2010: *Mobility and Social Organization on the Ancient Anatolian Black Sea Coast: An Archaeological, Spatial and Isotopic Investigation of the Cemetery at İkiztepe, Turkey*. PhD thesis, University of Toronto

Whittle, A.W., Bickle, P. (eds) 2014: *Early Farmers: The View from Archaeology and Science*. Oxford, Oxford University Press

Wood, A.R., Esko, T., Yang, J., Vedantam, S., Pers, T.H., Gustafsson, S., Lichtner, P. 2014: 'Defining the role of common variation in the genomic and biological architecture of adult human height' *Nature Genetics* 46.11: 1173–86

Yılmaz, H., Pehlevan, C. 2015: 'Van Çatak Erken Demir Çağ iskeletlerinde diş çürüğü' *Höyük* 8: 17–57

9. Diachronic changes in the frequency of caries in Anatolian past populations

M. Melis Koruyucu

Introduction

Data obtained from dental pathology can be crucially important for reconstructing the lifestyles of past people, as well as understanding their processes of adaptation to the environment (Kelley et al. 1991; Lieverse et al. 2007; Nicklisch et al. 2016; Lukacs 2017; Munoz 2017; Masotti et al. 2019). Since teeth are in direct contact with food and the environment, a strong relationship has been established between dental diseases and dietary habits, food preparation techniques and cultural characteristics of populations (Hillson 2008; Lukacs 2012). Therefore, the examination of dental remains can provide insight into the interactions between culture, ecological environment and the subsistence economies of past populations (Eshed et al. 2006; Lieverse et al. 2007; Lukacs 2012).

Dental caries is the most frequently used data source for reconstructing the dietary habits of past people, as it provides considerable information on the types of foods predominantly consumed. It is defined as the destruction of dental tissues due to continuous demineralisation caused by organic acids produced during the bacterial fermentation of carbohydrates (Lukacs 1989; Larsen et al. 1991; Hillson 2001; 2008; Lanfranco, Eggers 2010). The pH of dental plaque varies according to the amount of protein or carbohydrate intake. The bacteria in the dental plaque produce lactic acid when they metabolise carbohydrates, and conversely, when they metabolise protein the waste they produce increase the alkalinity of the oral environment (Hillson 1979). Therefore, the factor that determines whether dental caries will develop or not is the pH balance that changes in parallel with the foods eaten over the course of the individual's life. While the minerals of the enamel dissolve during acid periods, extra minerals are deposited on tooth surfaces in alkaline phases. If the acid periods are more numerous and last longer than the alkaline periods, the enamel starts to demineralise and dental caries occur (Hillson 1979). For this reason, a diet predominantly based on carbohydrates increases the possibility of caries occurring, while a protein-based diet decreases the risk (Nicklisch et al. 2016). This is why caries, among other dental pathologies, is generally the first source of data used to estimate the protein and carbohydrate balance in past diets (Larsen et al. 1991; Eshed et al. 2006; Lieverse et al. 2007; Polo-Cerdá et al. 2007; Hillson 2008; Lanfranco, Eggers 2010; Liu et al. 2010; Miller et al. 2014; Nicklisch et al. 2016; Lukacs 2017).

Diachronic change in the caries frequencies of past populations is a subject that has been frequently discussed (Kelley et al. 1991; Larsen et al. 1991; Lukacs 1992; Lieverse et al. 2007; Liu et al. 2010; Slaus et al. 2011; Miller et al. 2014; Nicklisch et al. 2016; Munoz 2017). Although the general view is that the frequency of caries increases over time (Kelley et al. 1991; Larsen et al. 1991; Lukacs 1992; Slaus et al. 2011; Munoz 2017), there are also some studies showing continuity (Eshed et al. 2006; Lieverse et al. 2007; Miller et al. 2014) or even a diachronic decrease (Liu et al. 2010; Nicklisch et al. 2016) in caries prevalence. Studies suggesting continuity or a diachronic decrease in the frequency of caries mostly focus on differences or similarities in the ecological environments, and accordingly in the subsistence economies (Lieverse et al. 2007; Liu et al. 2010; Miller et al. 2014) or the consumption of animal resources more over time (Nicklisch et al. 2016). However, in most such studies, it is emphasised that incidence of caries tends to increase following the adoption of agriculture (Turner II 1979; Kelley et al. 1991; Larsen et al. 1991; Lukacs 1992; 2017; Slaus et al. 2011; Munoz 2017). In these studies, which examine large sample groups dated to successive time periods, it has been found that the frequency of caries increased towards the later periods, in parallel with the spread of agriculture (see Turner II 1979; Kelley et al. 1991; Larsen et al. 1991).

There are very few studies on the diachronic changes in the frequency of caries in Anatolia. Although the caries incidences in Anatolian past populations have been mentioned in some studies (Yiğit et al. 2005; Koca et al. 2006; Erkman 2008; Erkman et al. 2008; Yaşar et al. 2008; Gözlük Kırmızıoğlu et al. 2009; 2010; Çırak et al. 2013; Sarı 2014; Yılmaz, Pehlevan 2015), these studies did not go beyond a comparison between populations, and are far from defining diachronic change in the prevalence of caries in Anatolia. When the caries frequencies of Anatolian past populations are examined closely, there are some fluctuations; that is, there are some increases during periods in which the caries rate is expected to be lower, or decreases in the periods when it is anticipated to be higher. However, it is noteworthy that the possible causes of such findings are not addressed in detail in these studies. With that in mind, the aim of this study is to examine how the caries rate changed from the Pre-Pottery Neolithic to the Modern Age in Anatolia, and to try to understand the possible reasons behind the aforementioned fluctuations observed in caries frequency through time.

Materials and methods

The dental caries data examined within the scope of this study were obtained from published and accessible studies on Anatolian past populations (see Tab. 9.1, Fig. 9.1). The caries frequencies of a total of 60 skeletal populations were collated representing different regions of Anatolia and dated to a wide range of time periods, from the Pre-Pottery Neolithic to the Modern Age. This study includes populations in which only permanent teeth were examined for caries, and the frequency of caries was calculated based on the tooth count rather than the individual count. The number of analysed teeth, and the number and frequency of carious teeth, were noted for each population. However, as for the studies in which the numbers of analysed teeth and carious lesions are not present, only frequencies were used. In order to gain a better understanding of the diachronic changes, populations (in which the data for the number of analysed teeth and number of carious teeth are present) were pooled according to their periods, and the frequency of caries was recalculated only by period (see Fig. 9.2).

Results

On examining the caries frequencies of the examined Anatolian populations (Tab. 9.1, Fig. 9.1), it is observed that the overall frequency ranges from 0% (Körtik Tepe, PPNA: Özbek 2005 and Ağızören; Middle Bronze Age: Yılmaz, Açıkkol 2003) to 20.8% (Datça/Burgaz, Roman period: Arıhan et al. 2010). When the populations are evaluated with regard to their chronological period, it can be seen that the frequency of caries varies from 0% to 11.6% in the Neolithic, 2.1% to 17.5% in the Early Bronze Age, 0% to 6.8% in the Middle Bronze Age, 2.6% to 11.3% in the Iron Age, 2.6% to 20.8% in Hellenistic-Roman period, 2.9% to 19% in the medieval period and 7.3% to 16.9% in the Modern Age. Even though there are some fluctuations, when considering the trend line, it was detected that the overall frequency of caries increased from the Pre-Pottery Neolithic to the Modern Age (Fig. 9.1).

This diachronic increase is more distinct when the populations are pooled according to their time periods, and the frequency of caries is considered only in terms of those periods (Fig. 9.2). While no caries were observed in the PPNA period, the frequency of caries is 4.1% in the PPNB. On the other hand, the prevalence is almost the same in the Pottery Neolithic (6.9%) and the Early Bronze Age (6.3%). Although there is a decrease in the caries incidence towards the Middle Bronze Age (5.6%) and the Iron Age (3.9%), the frequency increases again to 6.3% in the Hellenistic-Roman period. Furthermore, the increase in the frequency of caries continues in the medieval period (8.7%) and the Modern Age (9.8%) (Fig. 9.2).

Discussion

Various studies have revealed that the frequency of caries has increased with the adoption of agriculture, and that this increase continued over time (Turner 1979; Kelley et al. 1991; Lukacs 1992; 2017; Koca et al. 2006; Munoz 2017). However, some examples from different parts of the world have shown the opposite (Eshed et al. 2006; Lieverse et al. 2007; Liu et al. 2010; Miller et al. 2014; Nicklisch et al. 2016). Comparing the Bronze-Iron Age people of Xinjiang, located in the frontier areas of China, whilst the Neolithic people lived in the central areas, Liu et al. (2010) stated that the frequency of caries in the earlier period is between 6.8% and 8%. On the other hand, the caries frequency of the Bronze-Iron Age sample is 2.5%. This difference was interpreted as suggesting that the populations differ from each other in terms of their exposure to cariogenic foods. Namely, living in an ecologically harsh environment in the frontier areas, the Bronze-Iron Age people mainly relied on a hunter-gatherer economy, whereas the Neolithic people in the central areas were engaged more in agriculture. In other words, different subsistence economies due to the different ecological environments played an important role in shaping the difference between their caries frequencies (Liu et al. 2010). Nicklisch et al. (2016) conducted a similar study in which 21 sites from central Germany dated to the Early, Middle, Late Neolithic and Early Bronze Age were analysed in order to understand the diachronic change in caries frequency. On examining the caries frequency, calculated according to the individual count, they observed a significant decrease from the Early Neolithic (54.8%) to the Late Neolithic (38.3%). In the Early Bronze Age, the frequency decreased even more (to 35.6%), which was found to be statistically significant. Moreover, stable isotope analyses carried out on the same populations suggest that whereas $\delta^{15}N$ isotope ratios were relatively low in the Early Neolithic, a continuous increase was observed from the Middle Neolithic onwards. This increase was related to the greater consumption of meat and dairy products in the later periods. Therefore, dietary habits changed significantly from the Early Neolithic to the Early Bronze Age, contrary to the general view, and this shift occurred due to an increase in protein consumption. According to the results of this study, while cereals played an important role in the life of early agriculturalists, animal sources were consumed more over time, which had a positive effect on oral and dental health (Nicklisch et al. 2016).

Besides the studies suggesting a diachronic decrease in the frequency of caries, there are also some studies proposing that no temporal change was observed in the caries incidence (Eshed et al. 2006; Lieverse et al. 2007; Miller et al. 2014). For instance, examining the skeletal populations of Natufian hunter-gatherers (10,500–8300 BC) and Neolithic populations (8300–5500 BC) from the southern Levant, Eshed et al. (2006) stated that both populations manifested similar rates of caries (6.4% and 6.7%, respectively). According to this study, an increased consumption of agricultural products rich in carbohydrates in the Neolithic period was not reflected in the caries frequencies, and the change in the subsistence economies in the Levant did not cause changes in dental health, as previously believed. On the other hand, analysing two distinct cultural groups dated to the Late Mesolithic-Early Neolithic and the Middle Neolithic-Early Bronze Age, respectively, in the Cis-Baikal region of Siberia, Lieverse et al. (2007) stated that the frequency of caries is less than 1% in each settlement, and there is no diachronic change in caries frequency. Since all sites were located on the banks of Lake Baikal or the Angara River, it was concluded that their diets pointed to high-protein content, including fish and/or terrestrial mammals and seals rather than foods containing carbohydrates. In this context, it was again emphasised that geographical or environmental factors play a crucial role in determining dietary habits (Lieverse et al. 2007). A similar study was conducted on two populations dated to the Middle and Late Bronze Ages in Kazakhstan (Miller et al. 2014). These two successive periods reflect a transition in the patterns of settlements, the demographic size of settlements and mortuary rituals. Although the archaeological record indicates significant changes, only a slight shift in dental health was observed between these two periods. While there was relative continuity between the two periods in terms of dental pathological lesions, it was found that the frequency of some diseases, such as caries, decreases a bit towards the late period. The frequency of caries at Bestamak dated to the Middle Bronze Age is 0.1%, whereas no carious lesions were detected at Lisakovsk (dated to the Late Bronze Age), which suggests that the diets of both populations were

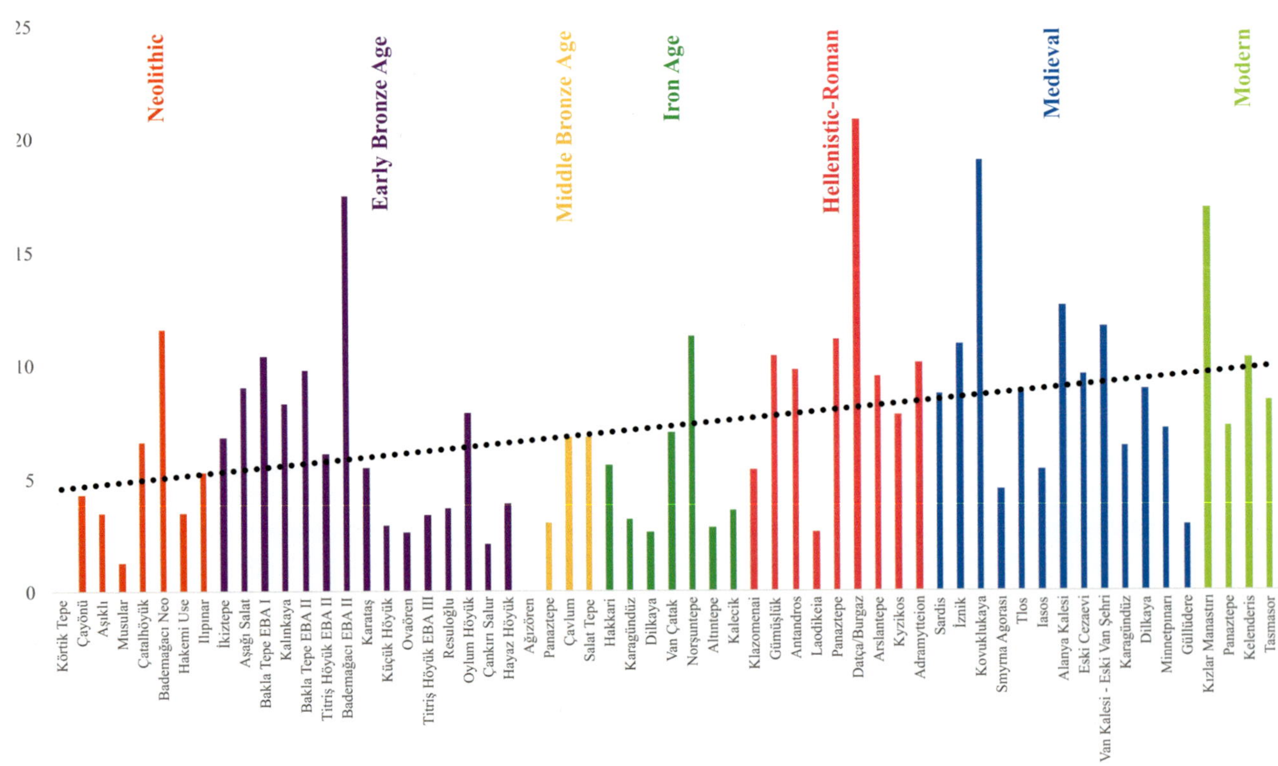

Fig. 9.1. Caries frequencies of Anatolian populations by time period (prepared by author).

Population	Period	N	n	%	Dental wear scores	Reference
Kortik Tepe	Neolithic	69	0	0.0	-	Özbek 2005
Çayönü	Neolithic	1945	84	4.3	5	Özbek 1997
Aşıklı	Neolithic	567	20	3.5	-	Özbek 2007
Musular	Neolithic	75	1	1.3	-	Özbek 2006
Çatalhöyük	Neolithic	485	32	6.6	-	Molleson et al. 2006
Bademağacı	Neolithic	-	-	11.6	5 and 6	Erdal 2009
Hakemi Use	Neolithic	342	12	3.5	-	Erdal 2013
Ilıpınar	Neolithic	399	21	5.3	-	Alpaslan-Roodenberg 2008
İkiztepe	Late Chalcolithic	3685	252	6.8	3.14*	Koruyucu 2019
Aşağı Salat	Late Uruk-Early Bronze Age	22	2	9.0	4	Başoğlu et al. 2013
Bakla Tepe	Early Bronze Age I	356	37	10.4	2.83*	Koruyucu 2019
Kalınkaya	Early Bronze Age I	-	-	8.3	-	Angel, Bisel 1986
Bakla Tepe	Early Bronze Age II	571	56	9.8	2.99*	Koruyucu 2019
Titris Höyük	Early Bronze Age II	277	17	6.1	2.79*	Koruyucu 2019
Bademağacı	Early Bronze Age II	97	17	17.5	2.82*	Koruyucu 2019
Karataş	Early Bronze Age II	-	-	5.5	-	Angel, Bisel 1986
Küçük Höyük	Early Bronze Age II	411	12	2.9	4	Açıkkol 2000
Ovaören	Early Bronze Age II	115	3	2.6	2	Başoğlu, Şener 2015
Titris Höyük	Early Bronze Age III	624	21	3.4	2.87*	Koruyucu 2019
Resuloğlu	Early Bronze Age III	1227	46	3.7	3.75*	Atamtürk, Duyar 2010
Oylum Höyük	Early Bronze Age III	642	51	7.9	3.04*	Sarı 2014
Çankırı Salur	Early Bronze Age III	48	1	2.1	4	Yiğit et al. 2011
Hayaz Höyük	Early Bronze Age III	229	9	3.9	1 and 2	Özbek 1984
Ağızören	Middle Bronze Age	62	0	0.0	4	Yılmaz, Açıkkol 2003
Panaztepe	Middle Bronze Age	299	9	3.0	3 and 4	Güleç, Duyar 1998
Çavlum	Middle Bronze Age	867	59	6.8	4	Sevim et al. 2005
Salat Tepe	Middle Bronze Age	118	8	6.8	3 and 4	Başoğlu et al. 2015
Hakkari	Iron Age	809	45	5.6	4	Gözlük et al. 2003
Karagündüz	Iron Age	1682	54	3.2	4	Erkman et al. 2008
Dilkaya	Iron Age	384	10	2.6	4	Erkman 2008
Van Çatak	Iron Age	114	8	7.0	-	Yılmaz, Pehlevan 2015
Norşuntepe	Iron Age	-	-	11.3	-	Korkmaz 1993
Altıntepe	Iron Age	808	23	2.8	2 and 3	Yiğit et al. 2005
Kalecik	Iron Age	165	6	3.6	4	Yılmaz et al. 2010
Klazomenai	7th–4th century BC	-	-	5.4	3	Yılmaz et al 2010
Gümüşlük	Classical-Hellenistic	-	-	10.4	-	Sağır et al. 2010
Antandros	7th–2nd century BC	420	41	9.8	4	Erdal 2000
Laodikeia	Roman	72	8	11.1	1 and 2	Güleç, Duyar 1998
Datça/Burgaz	Roman	24	5	20.8	3 and 4	Arıhan et al. 2010
Arslantepe	Late Roman	504	48	9.5	1 and 2	Uzel et al. 1988
Kyzikos	2nd century AD	-	-	7.8	2	Gözlük Kırmızıoğlu et al. 2009
Adramytteion	5th–6th century AD	208	21	10.1	4.04*	Atamtürk, Duyar 2008

Tab. 9.1 (cont. on next page).

Population	*Period*	*N*	*n*	*%*	*Dental wear scores*	*Reference*
Sardis	Late Roman-Early Byzantine	2114	183	8.7	2.94*	Eroğlu 1998
İznik	Medieval	5709	621	10.9	2.86*	Erdal 1996
Kovuklukaya	Medieval	179	8	4.5	-	cited in Erdal 2011
Smyrna Agorası	Medieval	-	-	19.0	-	Yaşar et al. 2008
Tlos	Medieval	-	-	8.9	-	Atamtürk et al. 2017
Iasos	Medieval	1374	74	5.4	3	Usta 2013
Alanya Kalesi	Medieval	246	31	12.6	-	Üstündağ, Demirel 2009
Eski Cezaevi	Medieval	469	45	9.6	-	Erdal 2003
Van Kalesi-Eski Van Şehri	Medieval	766	90	11.7	4	Gözlük et al. 2004
Karagündüz	Medieval	2734	174	6.4	4	Gözlük 2004
Dilkaya	Medieval	1964	174	8.9	3	Erkman 2008
Minnetpınarı	Medieval	1160	83	7.2	4	Yaşar 2007
Güllüdere	Medieval	376	11	2.9	-	Yaşar 2007
Kızlar Manastırı	15th century AD	409	69	16.9	2	Gözlük Kırmızıoğlu et al. 2010
Panaztepe	Islamic	-	-	7.3	3	Güleç 1989
Kelenderis	Modern	717	74	10.3	3	Çırak et al. 2013
Tasmasor	Modern	2329	196	8.4	4	Erdal 2011

*Tab. 9.1. Caries frequencies and dental wear scores of Anatolian populations. **N**: Number of observed teeth; **n**: Number of carious teeth; **%**: frequency of caries. Dental wear scores with * are the mean values of dental wear while the others are the most frequent scores.*

rich in protein. Consistent with the stable carbon and nitrogen isotopic data, these findings show that the subsistence strategy of these populations were based on pastoralism or hunting and gathering (Miller et al. 2014).

Whilst some of the previously discussed studies suggest a diachronic decrease or a continuity in the frequency of caries, it can be said that the majority of studies propose a diachronic increase in the caries rate (Turner 1979; Kelley et al. 1991; Larsen et al. 1991; Lukacs 1992; 2017; Koca et al. 2006; Slaus et al. 2011; Munoz 2017). Evaluating various research within his study, Turner (1979) reported that the prevalence of caries is 1.3% in hunter-gatherers, 4.84% in populations that relied on mixed economies and 10.43% in agricultural groups. In another study carried out on ancient populations from the Georgia and Florida coasts in the USA between 1000 BC and AD 1702, the sample populations were divided into four groups before and after European contact, and it was found that the frequency of caries increased over time (Larsen et al. 1991). While the frequency of caries in individuals from the precontact preagricultural group is 9%, this frequency reaches 59% in precontact agriculturalist and 82% in the late contact group. These significant leaps were interpreted as a result of the adoption and intensification of maize agriculture and the consumption of maize in greater quantities (Larsen et al. 1991). Similarly, examining five populations from hunter-gatherer-fishermen to those engaged in intensive agriculture, Kelley et al. (1991) argued that the frequency of caries increased in parallel with the subsistence economies of those five populations. Namely, the frequency of caries, which is 0.6–2.5% in hunter-gatherer-fishermen, increased to 11.5% in incipient agriculturalists, and to 14.4–48.1% in intensive agriculturalists (Kelley et al. 1991). Moreover, according to Lukacs (1992), who suggested that the caries frequency in South Asia increased in the transition from hunter-gatherer to agriculture, the caries incidence is 0–1.2% in hunter-gatherers of the Mesolithic period, 1.8–6.8% in the Bronze Age, and 4.4–7.7% in the Iron Age. These studies, in which considerably large sample groups were examined, are important in terms of demonstrating the increase in frequency of caries from the beginning to the wider spread and adoption of agriculture.

Can sample size affect caries rate?

It can be stated that the data obtained from these studies emphasising a diachronic increase in the frequency of caries fits with the data obtained from Anatolian populations. However, when examining Figure 9.1, there should be some factors that can further explain sudden or unexpected increases and decreases. One of these factors may be related to the sample size. For example, in some populations such as Körtik Tepe (Özbek 2005), Musular (Özbek 2006), Aşağı Salat (Başoğlu et al. 2013), Bademağacı (Koruyucu 2019), Ovaören (Başoğlu, Şener 2015), Çankırı Salur (Yiğit et al. 2011), Ağızören (Yılmaz, Açıkkol 2003), Salat Tepe (Başoğlu et al. 2015), Van Çatak (Yılmaz, Pehlevan 2015), Roman period Panaztepe (Güleç, Duyar 1998) and Datça/Burgaz (Arıhan et al. 2010), the number of analysed teeth is so limited that it does not exceed the total number of teeth for three individuals. Moreover, in some of those populations, the number of analysed teeth is even less than the normal number of teeth in one mouth. For instance, the number of analysed teeth is 22 at Aşağı Salat, and two of them have dental caries (Başoğlu et al. 2013; see Tab. 9.1). In this case, the frequency of caries is 9%, which cannot be considered to be a low incidence. Furthermore, the frequency of caries found on a total of 24 teeth analysed in Datça/Burgaz is 20.8% (n=5; Arıhan et al. 2010), which is the highest caries frequency among all Anatolian populations (see Fig. 9.1). As can be seen in these examples, even if the number of carious teeth is low, it appears that the frequency of caries is very high because the total number of teeth is low. Indeed, similar examples can be observed in Table 9.1. Therefore, it should be acknowledged that the sample size may cause problems in terms of reflecting the population, and may lead to some misleading results regarding the caries frequency.

Age-at-death as a factor in the frequency of caries

Another factor that is effective in accounting for the fluctuating patterns in Figure 9.1 is age. It has been established that there is a strong positive correlation between caries frequency and age (Hillson 2001). In parallel with the time the teeth spend in the mouth, the frequency of caries increases from young to old adulthood. For this reason, it is expected that the frequency of caries will be higher in populations where old adults are in the majority or the mean age at death is high. If we evaluate the Anatolian populations from this perspective, it will be more appropriate to give examples from the populations that lived in the same period because examples from populations dated to different periods may point to factors other than age. For instance, when considering İznik (Erdal 1996) and Kovuklukaya

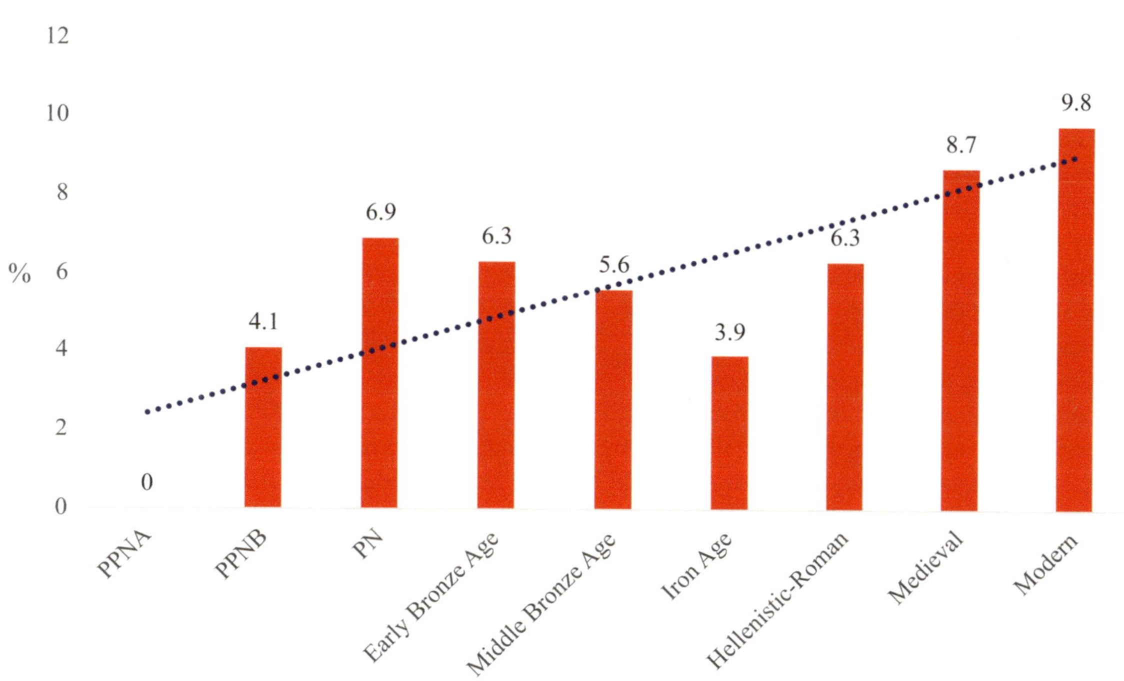

Fig. 9.2. Diachronic changes in the frequency of caries in Anatolian populations (prepared by author).

(cited in Erdal 2011) populations dated to the medieval period, it is observed that the frequency of caries is 10.9% and 19%, respectively (see Tab. 9.1). Even when the corrected caries frequencies are examined, the difference between İznik (14.9%; Duyar, Erdal 2003) and Kovuklukaya (27.2%; Erdal, unpublished data) is approximately doubled. The same situation arises when the two samples are compared in terms of mean age at death. The mean age at death in the Kovuklukaya sample population ($\bar{x}$=46.3; Erdal 2004) is almost twice the mean age at death at İznik ($\bar{x}$=25.6; Erdal 1996). Furthermore, more than half of the Kovuklukaya sample population is represented by old adults (55.6%; Erdal 2004), while a significant majority of the İznik population consists of young adults (Erdal 1996). On the other hand, antemortem tooth loss, which is a common condition in old age, may cause the caries frequency to be found relatively low in this age group. In addition, severe dental wear with increasing age can mask the true frequency of caries leading to an observed lower incidence, whilst a slow progression of dental wear may increase the observed caries incidence (Hillson 2001). Thus, the age factor and caries appear to have a complex relationship involving both dental wear and antemortem tooth loss. When comparing the caries frequencies of ancient human populations, evaluating such factors is important for providing a better understanding of possible temporal changes in the caries rate.

Caries versus dental wear

When diachronic change in the frequency of caries in Anatolia is considered, another and possibly the most important factor is dental wear. Previous studies have shown that the severity and speed of dental wear decreases in a diachronic manner, in terms of archaeological/chronological period (Smith 1984; Eshed et al. 2006; Deter 2009; Sołtysiak 2011; Tomczyk, Zalewska 2016). Due to the differences in food preparation techniques and daily activities, dental wear in hunter-gatherers is faster and more severe than in agriculturalists. One explanation is that hunter-gatherers consumed fibrous, hard and coarse foods. Additionally, the use of stone mortars and pestles for food processing may have introduced contaminants such as stone fragments and sand, which can easily damage dental tissues. However, with the onset of agriculture not only the types of food, but also the preparation techniques changed. That is to say, earthen vessels (i.e. pottery) replaced those made from stone, and the use of pottery to boil or heat foods provided a softer diet, which results in a decrease in the dental wear observed in agriculturalists compared to hunter-gatherers (Smith 1984; Eshed et al. 2006; Deter 2009). Comparing the Natufian hunter-gatherers and Neolithic populations, Eshed et al. (2006) suggest that changes in food preparation techniques and therefore dental wear may help to explain the differences in tooth condition in populations before and after the Neolithic agricultural revolution. It has been stated that the mean value of dental wear, which was 6.6–6.7 in the Natufian period, decreased to 6.1–6.3 in the Neolithic period (Eshed et al. 2006). Likewise, in another study examining the populations inhabiting the Syrian Lower Euphrates valley and dated to six different periods from the Early Bronze Age to the modern Islamic period, it has been observed that the diet was harder, poorly processed and more contaminated in the Early and Middle Bronze Ages. In contrast, the foods were free from foreign substances, well processed and softer in the later chronological periods (Tomczyk, Zalewska 2016). Despite the limited sample size, in his study on two archaeological sites in the Middle Euphrates valley of Syria, Sołtysiak (2011) discovered that dental wear decreases over time. In parallel to the grain milling technology that developed with the introduction of large rotary querns and watermills, it was determined that marks of dental microwear such as lines, pits and punctures observed in the people of the Late Roman/Islamic period were significantly less frequent than those from the Bronze Age (Sołtysiak 2011). A similar study on dental microwear was carried out on Neolithic and Early Bronze Age populations in Anatolia (Boz 2006). Comparing the dental microwear of the Neolithic Aşıklı and Çatalhöyük populations with the Early Bronze Age populations of İkiztepe and Oylum Höyük, Boz (2006) states that Neolithic people consumed harder, coarser and more poorly processed foods than Early Bronze Age people.

The data obtained from dental microwear is compatible with the dental wear data. That is to say, the most common degrees of dental wear in Neolithic populations are '5' and '6' (see Tab. 9.1), and therefore, it can be stated that the diet of this period was quite abrasive (Özbek 1997; Alpaslan-Roodenberg 2008; Erdal 2009; 2013). On the other hand, it is observed that the mean value of dental wear or the most common degree is not above '4' in any of the Early Bronze Age populations (Özbek 1984; Açıkkol 2000; Atamtürk, Duyar 2010; Yiğit et al. 2011; Başoğlu et al. 2013; Sarı 2014; Başoğlu, Şener 2015; Koruyucu 2019). This data has been associated with the consumption of well processed and non-abrasive foods in many populations (Özbek 1984; Sarı 2014; Başoğlu, Şener 2015; Koruyucu 2019). It can also be said that in the Middle Bronze Age and Iron Age the dental wear did not differ much from the Early Bronze Age (see Tab. 9.1). As for the Roman period, the dental wear is relatively mild, and the most common degrees of wear are '1–2' and '3–4' (Uzel et al. 1988; Güleç, Duyar 1998; Arıhan et al. 2010; Şimşek 2011). Similarly, '2', '3' and '4' are the also the most frequent scores in the medieval period (Erdal 1996; Gözlük 2004; Gözlük Kırmızıoğlu et al. 2004; 2010; Yaşar 2007; Erkman 2008; Usta 2013) and the Modern Age (Güleç 1989; Erdal 2011; Çırak et al. 2013). Therefore, it is noteworthy that there is a considerable decrease in dental wear from the Neolithic period to the Modern Age. As is well known, due to their sticky properties, well-processed and soft foods are more cariogenic than poorly ground or coarse-grained foods (Turner 1979; Hillson 2008). Accordingly, consumption of relatively refined, soft and small-grained foods due to the developments in food preparation techniques resulted in a decrease in dental wear on one hand, and an increase in the frequency of caries on the other. However, there has currently been no comprehensive study on the relation between caries and dental wear in Anatolia. Instead, studies have dealt with these forms of data independently. Thus, a negative or positive correlation between caries and dental wear should be analysed in more detail for each period.

Consequently, when evaluating the time periods in general, while no dental caries were observed in pre-agricultural populations, caries incidence suddenly increased with the transition to agriculture in the PPNB. This increase continues in the Pottery Neolithic period, which should be linked to both agriculture becoming more widespread as a subsistence economy, and to advances in food preparation techniques such as cooking and boiling foods in earthenware (Smith 1984; Eshed et al. 2006; Deter 2009). The frequency of caries in the Early Bronze Age is almost the same as in the Pottery Neolithic period. The Early Bronze Age is defined as a period during which many radical changes occurred in the social, political and economic structures of populations (Frangipane 2002; Sagona, Zimansky 2015; Massa 2016). However, it can be said that these radical changes were not reflected in dietary patterns, and the subsistence economy was based on agriculture and animal husbandry, as in the Pottery Neolithic period. Moreover, it is stated that the mean values of $\delta^{13}C$ and $\delta^{15}N$ are quite similar to each other in the Neolithic and Early Bronze Age (see Irvine, this volume). Although the frequency of caries seems to decrease in the Middle Bronze Age and Iron Age, this may be related to sample size. As a matter of fact, the number of sites dated to these periods is limited, and the number of teeth recovered from these sites is also small (Tab. 9.1). The caries incidence increases regularly again from the Hellenistic-Roman period to the Modern Age, which can be interpreted as an indicator of a more concentrated focus on arable agriculture. However, especially with the Roman period, the diversification of food choices and therefore the increasing possibility of consuming more cariogenic foods (Wilkins, Hill 2006; Killgrove, Tykot 2013; Murphy 2016) should also be effective in explaining the increase in the frequency of caries.

Conclusion

Due to the sample sizes and the different methodologies on which the studies are based, including which carious teeth were or were not considered for the purposes of the study, there are several critical handicaps when evaluating the collation of caries data collected by many different researchers. Therefore, data collected by the same researchers and on larger sample groups with similar techniques will allow us to better understand the temporal change in the frequency of caries and its possible causes. However, the currently available data does allow us to draw some inferences.

On examining the diachronic changes in the frequency of caries in ancient Anatolian populations, it is possible to say that there is an increase from the Pre-Pottery Neolithic period to the Modern Age. However, this increase follows a fluctuating pattern rather than a regular and continuous increase. These fluctuations are observed both between periods and between populations dated to the same period, and may be related to many different factors. Namely, the sample size of some populations is so small that interpretations may be limited and misleading. Furthermore, as age has a cumulative effect on dental caries, the frequency may be observed to be higher than it actually is in populations with a high mean age at death.

In addition, perhaps the most important issue to be considered when evaluating the diachronic changes in the frequency of caries is dental wear. Due to the developments in food preparation techniques, dental wear decreased over time. In contrast, the consumption of well-processed, soft and adhesive foods increased the caries incidence. Therefore, this inverse relationship between caries and dental wear should also be taken into account when considering the change in frequency of caries over time.

On the other hand, one of the most important reasons for the increase in caries frequency is an increase in carbohydrate consumption. This can be demonstrated in this study by observing that the largest increase in caries frequency in Anatolian populations occurred with the transition to agriculture, as has been suggested and demonstrated for other populations from different parts of the world. As agriculture became more widespread as a subsistence economy, the frequency of caries continued to increase into the later chronological periods. In conclusion, both the increase in carbohydrate consumption with the adoption of agriculture, and the introduction of more adhesive and fine-grained foods because of improvements in food preparation techniques are effective for explaining the increase in the frequency of caries over time.

Bibliography

Açıkkol, A. 2000: *Küçükhöyük Eski Tunç Çağı İnsanlarının Paleoantropolojik Açıdan İncelenmesi*. Master's thesis, Ankara University, Ankara

Alpaslan-Roodenberg, S. 2008: 'The Neolithic cemetery: the anthropological view' in J. Roodenberg, S. Alpaslan-Roodenberg (eds), *Life and Death in a Prehistoric Settlement in Northwest Anatolia: The Ilıpınar Excavations, Volume III, with contributions on Hacılartepe and Menteşe*. Leiden, Instituut voor Het Babije Oosten: 35–68

Angel, J.L., Bisel, S.C. 1986: 'Health and stress in an Early Bronze Age population' in B.H. Fowler, W.G. Moon (eds), *Ancient Anatolia: Aspects of Change and Cultural Development*. Madison, University of Wisconsin Press: 12–31

Arıhan, S.K., Çırak, A., Erkman, A.C. 2010: 'Datça/Burgaz iskeletlerinin paleoantropolojik analizi' *Arkeometri Sonuçları Toplantısı* 25: 297–310

Atamtürk, D., Duyar, İ. 2008: 'Adramytteion (Örentepe) iskeletlerinde ağız ve diş sağlığı' *Hacettepe Üniversitesi Edebiyat Fakültesi Dergisi* 25: 1–15

— 2010: 'Resuloğlu Erken Tunç Çağı topluluğunda ağız ve diş sağlığı' *Hacettepe Üniversitesi Edebiyat Fakültesi Dergisi* 27: 33–52

Atamtürk, D., Duyar, İ., Özener, B., Şahin, A. 2017: 'Tlos Orta Bizans Dönemi iskeletlerinin ağız ve diş sağlığı yönünden incelenmesi' *Arkeometri Sonuçları Toplantısı* 32: 133–47

Başoğlu, O., Şener, T. 2015: 'Ovaören Erken Tunç Çağı insanları: antropolojik bir inceleme' *Hitit Üniversitesi Sosyal Bilimler Enstitüsü Dergisi* 8.2: 455–76

Başoğlu, O., Akçay, A., Gözlük Kırmızıoğlu, P., Gökkoyun, S., Şener, T. 2013: 'Diyarbakır/Aşağı Salat Höyüğü iskeletleri' *Olba* 21: 27–44

Başoğlu, O., Kırmızıoğlu, P.G., Pehlevan, C., Yiğit, A., Erkman, A.C. 2015: 'The people of Diyarbakır/Salat Tepe in the Chalcolithic and Middle Bronze Age' *Mediterranean Archaeology and Archaeometry* 15.3: 237–47

Boz, B. 2006: *Dişlerde Mikro Aşınma Tekniğinin Beslenme Alışkanlıklarını Belirlemedeki Yeri: Eski Anadolu Toplumlarından Örnekler*. PhD thesis, Hacettepe University, Ankara

Çırak, A., Çırak, M.T., Erkman, A.C. 2013: 'Kelenderis Halkının diş ve çene paleopatolojileri' *Olba* 21: 1–25

Deter, C.A. 2009: 'Gradients of occlusal wear in hunter-gatherers and agriculturalists' *American Journal of Physical Anthropology* 138: 247–54

Duyar, İ., Erdal, Y.S. 2003: 'A new approach for calibrating dental caries frequency of skeletal remains' *HOMO – Journal of Comparative Human Biology* 54.1: 57–70

Erdal, Y.S. 1996: *İznik Geç Bizans Dönemi İnsanlarının Çene ve Dişlerinin Antropolojik Açıdan İncelenmesi*. PhD thesis, Hacettepe University, Ankara

— 2000: 'Antandros insanlarında ağız sağlığı' *Türk Arkeoloji ve Etnografya Dergisi* 1: 45–55

— 2003: 'Büyük Saray-Eski Cezaevi çevresi kazılarında gün ışığına çıkarılan insan kalıntılarının antropolojik analizi' *Arkeometri Sonuçları Toplantısı* 18: 15–30

— 2004: 'Kovuklukaya (Boyabat, Sinop) insanlarının sağlık yapısı ve yaşam biçimleriyle ilişkisi' *Anadolu Araştırmaları* 17.2: 169–96

— 2009: 'Bademağacı Erken Neolitik insanları' *Arkeometri Sonuçları Toplantısı* 24: 97–118

— 2011: 'Tasmasor Yakınçağ nekropolü ve iskeletlerinin antropolojik açıdan değerlendirilmesi' in S.Y. Şenyurt (ed.), *Tasmasor*. Ankara, Bilgin Yayınları: 255–359

— 2013: 'Life and death at Hakemi Use' in O.P. Nieuwenhuyse, R. Bernbeck, P.M.M.G. Akkermans, J. Rogasch (eds), *Interpreting the Late Neolithic of Upper Mesopotamia*. Turnhout, Brepols Publishers: 213–23

Erkman, A.C. 2008: *Van Dilkaya Erken Demir Çağı ve Orta Çağ Toplumunda Ağız ve Diş Sağlığı*. PhD thesis, Ankara University, Ankara

Erkman, A.C., Şimşek, N., Çırak, A., Arıhan, S.K. 2008: 'Karagündüz Erken Demir Çağı Toplumunda ağız ve diş sağlığı' *Arkeometri Sonuçları Toplantısı* 23: 141–56

Eroğlu, S. 1998: *Sardis Roma-Bizans Toplumlarında Diş Hastalıkları ve Ağız Sağlığı*. Master's thesis, Ankara University, Ankara

Eshed, V., Gopher, A., Hershkovitz, I. 2006: 'Tooth wear and dental pathology at the advent of agriculture: new evidence from the Levant' *American Journal of Physical Anthropology* 130: 145–59
Frangipane, M. 2002: *Yakındoğu'da Devletin Doğuşu*. İstanbul, Arkeoloji ve Sanat Yayınları
Gözlük Kırmızıoğlu, P., Yaşar, Z.F., Yiğit, A., Alpaslan, F.S., Erol, A.S., Kesikçiler, B. 2010: 'Trabzon kızlar manastırı iskeletlerinde ağız ve diş sağlığı' *Arkeometri Sonuçları Toplantısı* 25: 127–50
Gözlük Kırmızıoğlu, P., Yaşar, Z.F., Yiğit, A., Erol, A.S. 2009: 'Kyzikos iskeletlerinin dental analizi' *Arkeometri Sonuçları Toplantısı* 24: 139–62
Gözlük, P. 2004: *Van-Karagündüz Populasyonunun Dişlerinin ve Çenelerinin Paleopatolojik Açıdan İncelenmesi*. PhD thesis, Ankara University, Ankara
Gözlük, P., Yiğit, A., Erkman, A.C. 2004: 'Van Kalesi ve Eski Van şehri insanlarındaki sağlık sorunları' *Arkeometri Sonuçları Toplantısı* 19: 51–62
Gözlük, P., Yılmaz, H., Yiğit, A., Açıkkol, A., Sevim, A. 2003: 'Hakkari Erken Demir Çağı iskeletlerinin paleoantropolojik açıdan incelenmesi' *Arkeometri Sonuçları Toplantısı* 18: 31–40
Güleç, E. 1989: 'Panaztepe iskeletlerinin paleoantropolojik ve paleopatolojik incelenmesi' *Türk Arkeoloji Dergisi* 28: 73–95
Güleç, E., Duyar, İ. 1998: 'Panaztepe MÖ İkinci Bin ve Roma Dönemi iskeletlerinin antropolojik analizi (1985–1990)' *Antropoloji* 13: 179–206
Güleç, E., Sevim, A., Özer, İ., Sağır, M. 1998: 'Klazomenai'de yaşamış insanların sağlık sorunları' *Arkeometri Sonuçları Toplantısı* 13: 133–59
Hillson, S. 1979: 'Diet and dental disease' *World Archaeology, Food and Nutrition* 11.2: 147–62
— 2001: 'Recording dental caries in archaeological human remains' *International Journal of Osteoarchaeology* 11: 249–89
— 2008: 'Dental pathology' in M.A. Katzenberg, S.R. Saunders (eds), *Biological Anthropology of the Human Skeleton*, second edition. New York, John Wiley & Sons, Inc.: 301–40
Kelley, M.A., Levesque, D.R., Weidl, E. 1991: 'Contrasting patterns of dental disease in five early Northern Chilean groups' in M.A. Kelley, C.S. Larsen (eds), *Advances in Dental Anthropology*. New York, Wiley-Liss: 203–13
Killgrove, K., Tykot, R.H. 2013: 'Food for Rome: a stable isotope investigation of diet in the Imperial period (1st–3rd centuries AD)' *Journal of Anthropological Archaeology* 32.1: 28–38
Koca, B., Güleç, E., Gültekin, T., Akın, G., Güngör, K., Brooks, S.L. 2006: 'Implications of dental caries in Anatolia: from hunting-gathering to the present' *Human Evolution* 21: 215–22
Korkmaz, D. 1993: *Elazığ/Norşuntepe Demir Çağı İskeletlerinde Diş Yapısı ve Hastalıkları*. Master's thesis, Ankara University, Ankara
Koruyucu, M.M. 2019: *Anadolu Erken Tunç Çağı Topluluklarında Ağız ve Diş Sağlığı*. PhD thesis, Hacettepe University, Ankara
Lanfranco, L.P., Eggers, S. 2010: 'The usefulness of caries frequency, depth, and location in determining cariogenicity and past subsistence: a test on early and later agriculturalists from the Peruvian coast' *American Journal of Physical Anthropology* 143: 75–91
Larsen, C.S., Shavit, R., Griffin, M. 1991: 'Dental caries evidence for dietary change: an archaeological context' in M.A. Kelley, C.S. Larsen (eds), *Advances in Dental Anthropology*. New York, Wiley-Liss: 179–202
Lieverse, A.R., Link, D.W., Bazaliiskiy, V.I., Goriunova, O.I., Weber, A.W. 2007: 'Dental health indicators of hunter–gatherer adaptation and cultural change in Siberia's Cis-Baikal' *American Journal of Physical Anthropology* 134.3: 323–39
Liu, W., Zhang, Q.C., Wu, X.J., Zhu, H. 2010: 'Tooth wear and dental pathology of the Bronze-Iron Age people in Xinjiang, Northwest China: implications for their diet and lifestyle' *HOMO – Journal of Comparative Human Biology* 61.2: 102–16
Lukacs, J.R. 1989: 'Dental paleopathology: methods for reconstructing health status and dietary patterns in prehistory' in M.Y. Iscan, K.A.R. Kennedy (eds), *Reconstruction of Life from the Skeleton*. New York, Alan R. Liss: 261–86
— 1992: 'Dental paleopathology and agricultural intensification in South Asia: new evidence from Bronze Age Harappa' *American Journal of Physical Anthropology* 87: 133–50
— 2012: 'Oral health in past populations: context, concepts and controversies' in A.L. Grauer (ed.), *A Companion to Paleopathology*. West Sussex, Wiley-Blackwell: 553–81
— 2017: 'Dental adaptations of Bronze Age Harappans: occlusal wear, crown size, and dental pathology' *International Journal of Paleopathology* 18: 69–81

Masotti, S., Varalli, A., Goude, G., Moggi-Cecchi, J., Gualdi-Russo, E. 2019: 'A combined analysis of dietary habits in the Bronze Age site of Ballabio (Northern Italy)' *Archaeological and Anthropological Sciences* 11: 1029–47

Massa, M. 2016: *Networks Before Empires: Cultural Transfers in West and Central Anatolia During the Early Bronze Age*. PhD thesis, University College London, London

Miller, A.V., Usmanova, E., Logvin, V., Kalieva, S., Shevnina, I., Logvin, A., Kolbina, A., Suslov, A. 2014: 'Dental health, diet, and social transformations in the Bronze Age: comparative analysis of pastoral populations in Northern Kazakhstan' *Quaternary International* 348: 130–46

Molleson, T., Andrews, P., Boz, B. 2006: 'Reconstruction of the Neolithic people of Çatalhöyük' in I. Hodder (ed.), *Inhabiting Çatalhöyük Reports from 1995–99 Seasons by the Members of Çatalhöyük Teams*. London, Short Run Press: 279–300

Munoz, O. 2017: 'Transition to agriculture in South-Eastern Arabia: insights from oral conditions' *American Journal of Physical Anthropology* 164.4: 702–19

Murphy, C. 2016: 'Finding millet in the Roman world' *Archaeological and Anthropological Sciences* 8: 65–78

Nicklisch, N., Ganslmeier, R., Siebert, A., Friederich, S., Meller, H., Alt, K.W. 2016: 'Holes in teeth – dental caries in Neolithic and Early Bronze Age populations in Central Germany' *Annals of Anatomy* 203: 90–99

Özbek, M. 1984: 'Etude anthropologique des restes humains de Hayaz Höyük' *Anatolica* 11: 155–69

— 1997: 'Çayönü tarım toplumunda diş sağlığı' *Türk Arkeoloji Dergisi* 31: 181–216

— 2005: 'Körtik Tepe'de insan sağlığı' *Arkeometri Sonuçları Toplantısı* 20: 41–52

— 2006: 'Musular Neolitik insanlarının antropolojik analizi' in B. Avunç (ed.), *Hayat Erkanal'a Armağan: Kültürlerin Yansıması*. İstanbul, Homer Kitabevi ve Yayıncılık: 564–70

— 2007: *Dişlerle Zamanda Yolculuk*. Ankara, Hacettepe Üniversitesi Yayınları

Polo-Cerdá, M., Romero, A., Casabó, J., De Juan, J. 2007: 'The Bronze Age burials from Cova Dels Blaus (Vall d' Uixó, Castelló, Spain): an approach to palaeodietary reconstruction through dental pathology, occlusal wear and buccal microwear patterns' *HOMO – Journal of Comparative Human Biology* 58.4: 297–307

Sagona, A., Zimansky, P. 2015: *Arkeolojik Veriler Işığında Türkiye'nin En Eski Kültürleri: MÖ 1.000.000–550*. İstanbul, Arkeoloji ve Sanat Yayınları

Sağır, M., Satar, Z., Özer, İ., Güleç, E. 2010: 'Gümüşlük-Milas iskeletlerinin ağız ve diş sağlığı' *Arkeometri Sonuçları Toplantısı* 25: 69–78

Sarı, İ. 2014: *Oylum Höyük Erken Tunç Çağı Toplumunda Ağız ve Diş Sağlığı*. Master's thesis, Cumhuriyet University, Sivas

Sevim, A., Yılmaz, H., Açıkkol, A. 2005: 'Çavlum iskeletlerinin paleoantropolojik analizi' in *I. Uluslararası Dünden Bugüne Eskişehir Sempozyumu–Siyasal, Ekonomik, Sosyal ve Kültürel Yapı*. Eskişehir, Anadolu Üniversitesi Basımevi: 337–56

Slaus, M., Bedic, Z., Sikanjic, P.R., Vodanovic, M., Kunic, A.D. 2011: 'Dental health at the transition from the Late Antique to the Early Medieval period on Croatia's Eastern Adriatic coast' *International Journal of Osteoarchaeology* 21: 577–90

Smith, B.H. 1984: 'Patterns of molar wear in hunter-gatherers and agriculturalists' *American Journal of Physical Anthropology* 63.1: 39–56

Sołtysiak, A. 2011: 'Cereal grinding technology in Ancient Mesopotamia: evidence from dental microwear' *Journal of Archaeological Science* 38.10: 2805–10

Şimşek, N. 2011: *Laodikeia Populasyonunun Diş ve Çenelerinin Paleopatolojik Açıdan İncelenmesi*. PhD thesis, Ankara University, Ankara

Tomczyk, J., Zalewska, M. 2016: 'Mechanical and chemical dental wear in historical population from the Syrian Lower Euphrates Valley' *Archives of Oral Biology* 62: 49–57

Turner II, C.G. 1979: 'Dental anthropological indications of agriculture among the Jomon people of Central Japan' *American Journal of Physical Anthropology* 51: 619–35

Usta, N.D.Y. 2013: 'Iasos (Bizans Dönemi) toplumunda ağız ve diş sağlığı' *Antropoloji* 25: 117–54

Üstündağ, H., Demirel, F.A. 2009: 'Alanya Kalesi iskelet topluluğunda ağız ve diş sağlığı' *Hacettepe Üniversitesi Edebiyat Fakültesi Dergisi* 26: 219–34

Uzel, İ., Alpagut, B., Kofoğlu, S. 1988: 'Arslantepe (Malatya) Geç Roma Dönemi iskeletlerinde diş çürüğü aşınmaları ve periodontal hastalıklar' *Arkeometri Sonuçları Toplantısı* 3: 31–54

Wilkins, J., Hill, S. 2009: *Food in the Ancient World*. New York, Wiley-Blackwell

Yaşar, Z.F. 2007: *Adli Dental Antropoloji: (Dental Antropoloji Açısından Minnetpınarı ve Güllüdere Toplumlarının Dişlerinin Karşılaştırmalı Analizi)*. PhD thesis, Ankara University, Ankara

Yaşar, Z.F., Yiğit, A., Kırmızıoğlu, P.G., Erol, A.S. 2008: 'Smyrna Agorası insanlarının ağız ve diş sağlığı' *Arkeometri Sonuçları Toplantısı* 23: 127–40

Yılmaz, H., Açıkkol, A. 2003: 'Kütahya Ağızören iskeletlerine ait dişlerin incelenmesi' *Antropoloji* 17: 71–108

Yılmaz, H., Baykara, İ., Baykara, D. 2010: 'Kalecik (Van) insanlarının ağız ve diş sağlığı' *Arkeometri Sonuçları Toplantısı* 25: 15–31

Yiğit, A., Gözlük, P., Erkman, A.C., Çırak, A., Şimşek, N. 2005: 'Altıntepe Urartu iskeletlerinin paleoantropolojik açıdan değerlendirilmesi' *Arkeometri Sonuçları Toplantısı* 20: 79–90

Yiğit, A., Gözlük Kırmızıoğlu, P., İbiş, R., Erol, A.S. 2011: 'Çankırı Salur Erken Tunç Dönemi insanları' *Arkeometri Sonuçları Toplantısı* 26: 273–91

Yılmaz, H., Pehlevan, C. 2015: 'Van Çatak Erken Demir Çağ iskeletlerinde diş çürüğü' *Höyük* 8: 17–57

10. An isotopic perspective of diachronic changes in dietary habits and subsistence practices in Anatolia

Benjamin Irvine

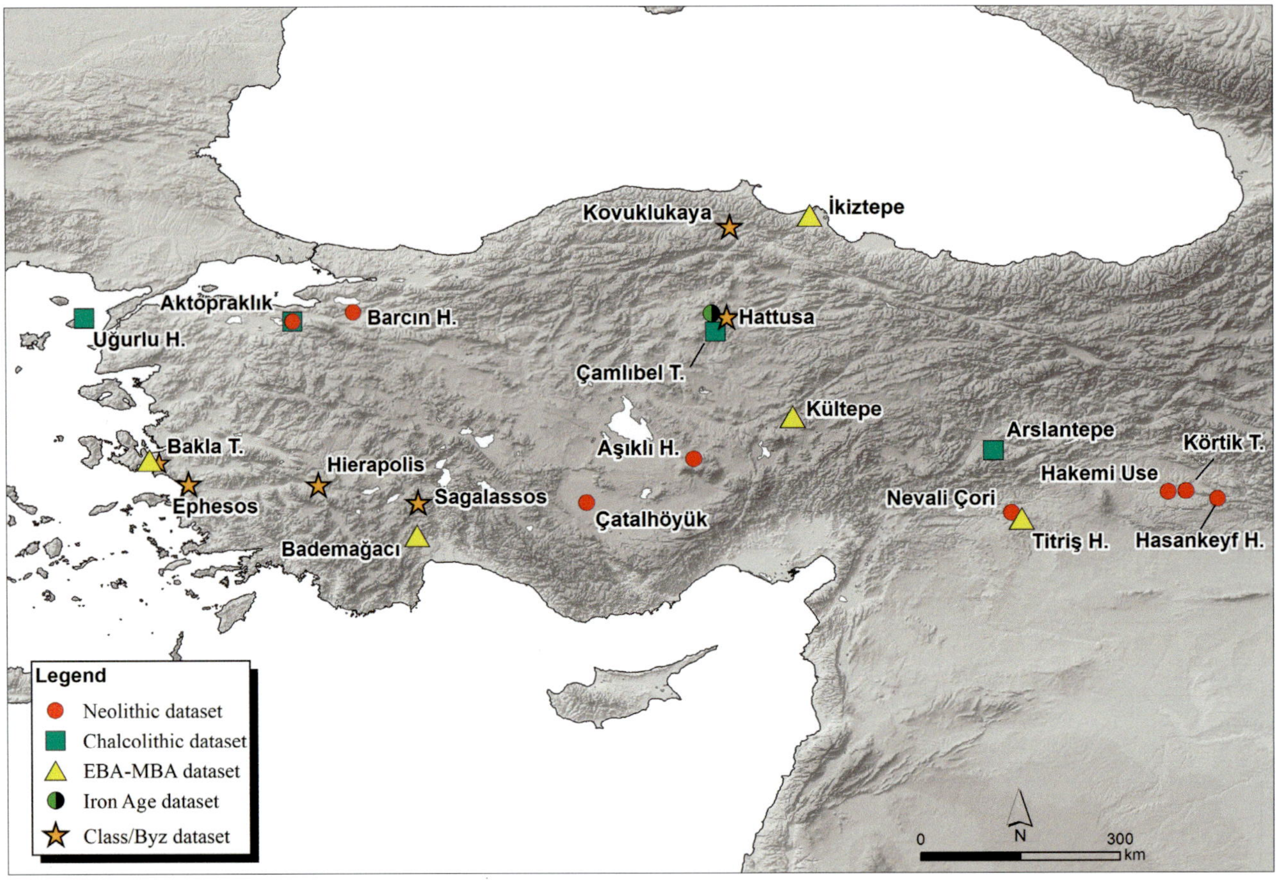

Fig. 10.1. Map of all sites in Anatolia with $\delta^{13}C$ and $\delta^{15}N$ stable isotope data (map by Michele Massa).

Introduction

This study provides a preliminary examination into diachronic changes in dietary habits, subsistence practices and agricultural strategies in the Anatolian Peninsula (hereafter Anatolia). This research does not provide a comprehensive and in-depth analysis, but rather it will provide a general overview with regard to discussing the most (current) clearly observable and identifiable large-scale patterns. For Anatolia there have been more detailed isotopic examinations (i.e. intra-population differences regarding finer scale chronological and demographic differences, such as between the sexes and ages) of individual site populations within their particular chronological timeframes (see references in Tab. 10.1, and also Irvine, Özdemir 2020 for an overview of isotopic research in Türkiye). There have been a few studies examining more specific multi-site isotopic patterns within, or across, particular time periods in Anatolia, such as for the Early Bronze Age (EBA) (Irvine et al. 2019), from the Neolithic to the end of the EBA (Irvine et al., forthcoming), and from the EBA to the Middle Byzantine period (Özdemir, Irvine, forthcoming). Outside Anatolia there have been other studies that have examined dietary habits and subsistence practices in a large-scale, regional and diachronic manner utilising stable isotopes. Dietary habits have been examined using stable isotopes of carbon (C), nitrogen (N) and sulphur (S) in the Central Balkans from the Mesolithic to the

Neolithic (Jovanović et al. 2019). Population dietary habits were examined from the Neolithic to the Iron Age of the Great Hungarian Plain utilising stable isotopes, in conjunction with genomic data (Gamarra et al. 2018). Papathanasiou (2015) and Papathanasiou and Richards (2015) examined long-term changes in dietary habits from the Neolithic to Byzantine periods of Greece by analysing C and N stable isotope data. To identify the introduction of millet into the south Caucasus, Herrscher et al. (2018) conducted a diachronic examination of populations from Europe to Asia using stable isotopes of C and N. Further afield from Anatolia and the Near East, Liu et al. (2020) examined long-term dietary changes across China utilising C and N stable isotope data. Pokutta (2017) examined dietary habits on a pan-regional rather than diachronic scale, examining $\delta^{13}C$ and $\delta^{15}N$ values from Bronze Age (third and second millennia BC) Iberia, Greece, Transcaucasia, Kazakhstan and China.

Site Name	*Location*	*Time Periods with $\delta^{13}C$ and $\delta^{15}N$ data*	*Reference*
Körtik Tepe*	Southeast Anatolia	Younger Dryas and Early Holocene (10,400–9200 cal. BC)	Benz et al. 2016
Hasankeyf Höyük	Southeast Anatolia	PPN (ca late 10th millennium BC)	Itahashi et al. 2017
Nevalı Çori	Southeast Anatolia	PPN/EN (ca 8720–7470 cal. BC)	Lösch et al. 2006
Aşıklı Höyük	Central Anatolia	PPN/EN (ca 8350–7300 cal. BC)	Itahashi et al. 2021
Barcın Höyük	Northwest Anatolia	LN (6600–6200 cal. BC)	Budd et al. 2020
Çatalhöyük	South Central Anatolia	LN (ca 8000–7000 bp)	Richards et al. 2003
Hakemi Use	Southeast Türkiye	LN (6100–5950 cal. BC)	Itahashi et al. 2019
Aktopraklık	Northwest Anatolia	LN to ECh (6400–5635 cal. BC)	Budd et al. 2013, 2018
Uğurlu	Gökçeada Island, Northeast Aegean	Ch (5500–4900 cal. BC)	Pilaar Birch et al. 2021
Arslantepe	East Anatolia	LCh (3800–3375 BC)	Iacumin et al. 2020
Çamlıbel Tarlası	North Central Anatolia	LCh (3650–3375 cal. BC)	Pickard et al. 2016
İkiztepe	North Anatolia	LCh-EBA I (ca late 4th millennium BC)	Irvine et al. 2019
Titriş Höyük	Southeast Anatolia	EBA I–III (3000–2100 BC)	Irvine et al. 2019
Bakla Tepe	Southwest Anatolia	EBA I and II/III (ca 3000–2800 and 2500–2300 BC), Late Roman/Early Byzantine	Irvine et al. 2019; Irvine unpublished data
Bademağacı	South Anatolia	EBA II (ca 2500 cal. BC)	Irvine et al. 2019
Kültepe-*Kanesh*	Central Anatolia	MBA (ca 1930–1720 BC)	Yazıcıoğlu-Santamaria 2015
Boğazköy/Hattuşa	North Central Anatolia	Iron Age (ca 10th–9th centuries BC), Hellenistic period (3rd century BC), Roman period (ca AD 2nd century)	Pickard et al. 2017
Sagalassos*	South/Southwest Anatolia	Classical-Hellenistic (400–200 BC), Late Imperial Roman period (AD 300–450), Mid. Byzantine (AD 800–1200)	Fuller et al. 2012
Ephesus	West Türkiye	Roman (AD 2nd and 3rd centuries)	Lösch et al. 2014
Hierapolis	West/West Central Anatolia	Roman/Early Byzantine (AD 1st–7th centuries), Mid. Byzantine (AD 9th–13th centuries)	Wong et al. 2017
Kovuklukaya	North Anatolia	Mid. Byzantine (ca cal. AD 770–970)	Özdemir 2018

Tab. 10.1. Summary overview of sites with stable isotope data ($\delta^{13}C$ and $\delta^{15}N$) used in this study. Time periods refer to local regional chronologies used by the researchers, where possible absolute dates have been included. PPN = Pre-Pottery Neolithic; LNeo = Late Neolithic; ECh = Early Chalcolithic; LCh = Late Chalcolithic; EBA = Early Bronze Age; MBA = Middle Bronze Age.

The use of quantifiable scientific data, like stable isotope signals of human and faunal osteological material, has been demonstrated to provide direct information, and therefore a more accurate understanding of dietary habits and subsistence strategies (Lee-Thorp 2008; Schwarcz, Schoeninger 2011; Larsen 2018). Isotopic studies of diet can reveal differences in dietary habits and access to food resources between individuals and populations, as well as diachronically (Quinn, Beck, 2016; Pokutta 2017). The most commonly utilised stable isotope ratios for examining dietary habits and associated subsistence strategies are C and N (Walker, DeNiro 1986; Faure, Mensing 2005; Pollard et al. 2007; Lee-Thorp 2008; Schwarz, Shoeninger 2011; Fernandes et al. 2012; Szpak et al. 2013). The background theory, method and application of stable isotope ratio analysis of C and N, from human and faunal bone collagen, and in order to reconstruct palaeodietary habits, has been well established and described elsewhere (see, amongst others, Schoeninger, DeNiro 1984; Walker, DeNiro 1986; Ambrose 1993; Ambrose, Norr 1993; Richards, Hedges 1999; Katzenberg 2000; Faure, Mensing 2005; Pollard et al. 2007; Lee-Thorp 2008; Schwarz, Shoeninger 2011).

Anatolia, due to its geographical location and orientation, has long acted as a non-passive and integrated conduit for material culture, ideas, agricultural practices (e.g. the Neolithic 'package'), (domestic) animals, genes and people (Zimmermann 2007; Şahoğlu 2008; Greenberg, Palumbi 2015; Heyd et al. 2016; Işıklı, Altunkaynak 2016; Massa, Palmisano 2018; Brami 2019; Skourtanioti et al. 2020). Ever since the advent of Neolithisation and sedentary domestic agriculturalism in the Fertile Crescent (Anatolia's southeast constituting the Crescent's northern boundary), Anatolia has been one of the key regions intrinsically connected with dietary, subsistence and agricultural developments. Therefore, a diachronic overview of the dietary habits and subsistence practices of ancient Anatolia is not only critical for gaining a better comprehension of its past peoples and their human-environment interactions, and social, economic and political dynamics, but also those of neighbouring regions and further afield.

Materials and methods

The stable isotope data was collected from, mostly, published research relating to 21 sites from time periods ranging from ca 10,000 BC to the 10th century AD, as well as from a variety of geographic and environmental regions of Anatolia (see Fig. 10.1; Tab. 10.1 provides a summary overview of the sample populations). The stable isotope values examined in this study constitute all of the currently available (as of December 2021) and accessible (δ^{13}C and δ^{15}N) data from Anatolia.

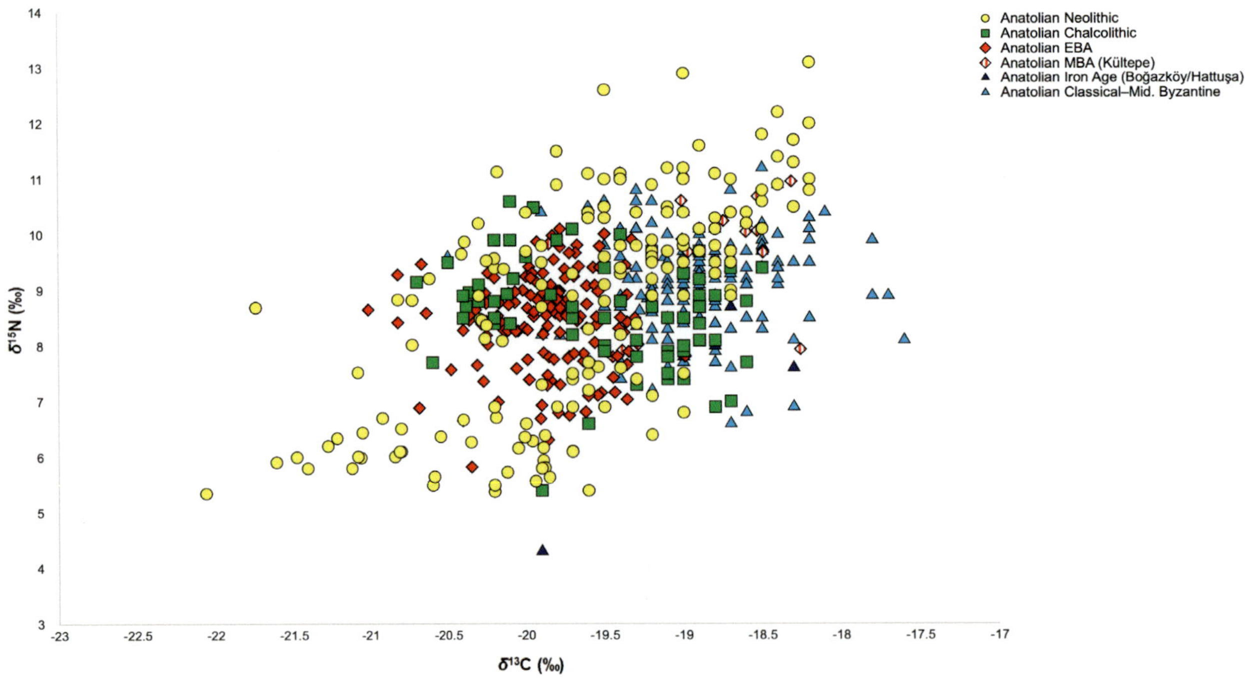

Fig. 10.2. δ^{13}C and δ^{15}N values for adults from Anatolian Neolithic to Byzantine period sample populations (chart by author).

Additionally, only $\delta^{13}C$ and $\delta^{15}N$ values obtained from the bulk bone collagen of adult human individuals was collated to enable comparability. Some $\delta^{13}C$ and $\delta^{15}N$ data from the publications was also excluded from examination in this study, if the collagen quality criteria (C/N ratio, %C, %N) was outside of the commonly accepted ranges (DeNiro 1985; Guiry, Szpak 2021), or if they were from individuals that could not be firmly identified as adults in age. Studies that reported only population means of $\delta^{13}C$ and $\delta^{15}N$ from sites (n=2 out of 21) were excluded from the statistical analyses and Figures 10.2 and 10.3, but have been noted in Table 10.1 (where these sites/populations are identified as having only population means) and Figure 10.1, and their means and standard deviations have been plotted in Figures 10.4 and 10.5.

The collated stable isotope data was then used to create means and standard deviations, as well as conducting further statistical analyses. A Shapiro-Wilks test was first employed to examine normality of distribution in the data. Following this, a Kruskal-Wallis ANOVA test, with a post-hoc Dunn's Procedure/two-tailed pairwise comparison test was performed. The statistical analyses were all performed using Excel with the XLSTAT add-on. The stable isotope data was also plotted to visualise diachronic changes and patterns (see Figs 10.2–10.5).

Results

A total of 521 data points (i.e. a $\delta^{13}C$ and $\delta^{15}N$ value), equating to roughly a similar number of individuals were collated from 19 sites (excluding the two sites that only have population means). To streamline the diachronic analysis of the data, the values from the respective site populations were combined according to their archaeologically determined time period, resulting in four chronological sample populations: the Neolithic, Chalcolithic, EBA and Classical to (Middle) Byzantine. The

Time Period	*$\delta^{13}C$ mean and SD (‰)*	*$\delta^{13}C$ Range with Minimum and Maximum Values (‰)*	*$\delta^{15}N$ mean and SD (‰)*	*$\delta^{15}N$ Range with Minimum and Maximum Values (‰)*
Neolithic (n=164)	-19.6 ± 0.8	3.9 (-22.1 to -18.2)	8.8 ± 2.0	7.8 (5.4 to 13.1)
Chalcolithic (n=74)	-19.5 ± 0.6	2.2 (-20.7 to -18.5)	8.6 ± 0.9	5.2 (5.4 to 10.6)
EBA (n=142)	-19.9 ± 0.3	2.0 (-21.0 to -19.0)	8.4 ± 0.8	4.2 (5.8 to 10.1)
Classical–Mid. Byzantine (n=120)	-18.9 ± 0.4	2.9 (-20.5 to -17.6)	9.1 ± 0.8	4.6 (6.6 to 11.2)

Tab. 10.2. Means, SD and ranges for $\delta^{13}C$ and $\delta^{15}N$ data from the Neolithic, Chalcolithic, EBA and Classical-Mid/Late Byzantine periods of Anatolia.

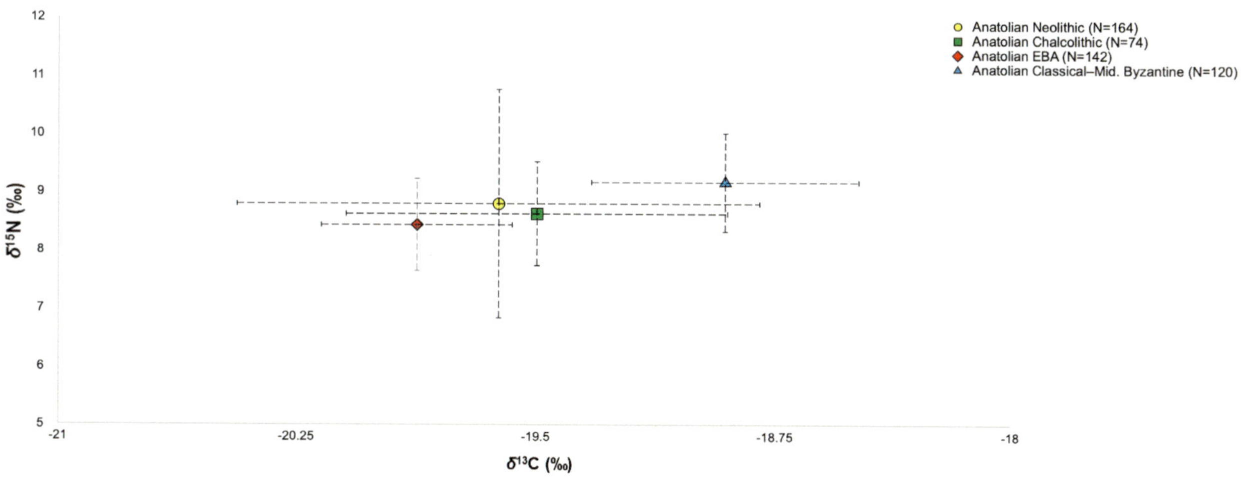

Fig. 10.3. Means and standard deviations of $\delta^{13}C$ and $\delta^{15}N$ values for adults from Anatolian Neolithic to Byzantine period sample populations (chart by author).

number of data points, means, standard deviations and ranges – including minimum and maximum values for each of these grouped sample populations – are summarised in Table 10.2. With the exception of the Chalcolithic period (64 data points from four sites), the MBA (15 data points from one site) and the Iron Age (six data points from one site) there is a relatively even and representative balance for the grouped time periods. As data for the MBA and Iron Age comes from only single sites, they are not included in Table 10.2, Figure 10.3 or the statistical analyses, the individual data points are present in Figure 10.2, and the means and standard deviations (SD) are included in Figures 10.4 and 10.5, however.

In general, it can be said that the $\delta^{13}C$ and $\delta^{15}N$ values from all time periods, and indeed the discrete populations, are indicative of a predominantly terrestrial C_3-based mixed diet (i.e. terrestrial plant and animal protein). However, an increase in the mean $\delta^{13}C$ values is noted in the Classical-Byzantine periods, which is likely suggestive of an increased C_4 component in the diet. Furthermore, there is a narrower range of $\delta^{13}C$ and $\delta^{15}N$ values in the EBA compared to earlier and later periods (see Tab. 10.2, Figs 10.2–10.4). The mean $\delta^{15}N$ values also increase in the historical periods (Tab. 10.2). Examining the data more closely (see Fig. 10.5), however, an increase in $\delta^{15}N$ values from the Early to later periods of the Neolithic is actually observable, followed by a reduction through the Chalcolithic and into the EBA, which is then followed by an increase in the historical periods. The range in $\delta^{15}N$ values is greatest for the (overall) Neolithic, then the Chalcolithic, and it is lowest in the EBA, and then, again, increases slightly in the historical periods. The implications of these results for reconstructing dietary habits and subsistence practices will be addressed in the discussion.

To further investigate potential differences between the time periods, statistical analyses were applied to the data; for all statistical tests $p<0.05$ was considered to be an indication of statistical significance. A Shapiro-Wilks test first examined the normality of distribution in the data. The only dataset to display normality were the $\delta^{15}N$ values for the Chalcolithic ($p=0.12$) and Classical-Middle Byzantine ($p=0.25$) periods. Therefore, a Kruskal-Wallis ANOVA test with a post-hoc Dunn's Procedure/two-tailed pairwise comparison test was performed. For $\delta^{13}C$, a strong overall statistical significance was determined between the sample populations ($p<0.0001$). The post-hoc pairwise comparison test found that there were statistically significant differences between the Neolithic and the EBA, the Neolithic and the Classical-Middle Byzantine, the Chalcolithic and the EBA and the Chalcolithic and the Classical-Middle Byzantine for $\delta^{13}C$ (see Tab. 10.3). For $\delta^{15}N$, the Kruskal-Wallis test determined that there was also an overall strong statistically significant difference between the sample populations ($p<0.0001$). The pairwise comparisons for $\delta^{15}N$ found that there were statistically significant differences between the Neolithic and EBA, the Chalcolithic and the Classical-Middle Byzantine and the EBA and Classical-Middle Byzantine (see Tab. 10.4). Therefore, conversely, no statistically significant differences were found between the other chronologically grouped sample populations, for either $\delta^{13}C$ or $\delta^{15}N$.

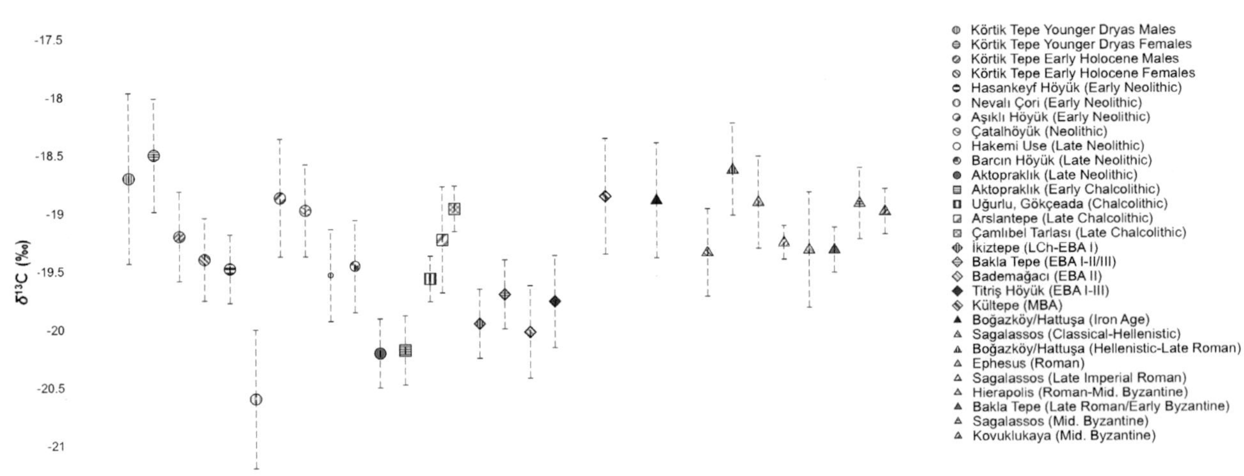

Fig. 10.4. Means and standard deviations for $\delta^{13}C$ values from Anatolian populations through time. For references and sources of the data see Table 10.1 (chart by author).

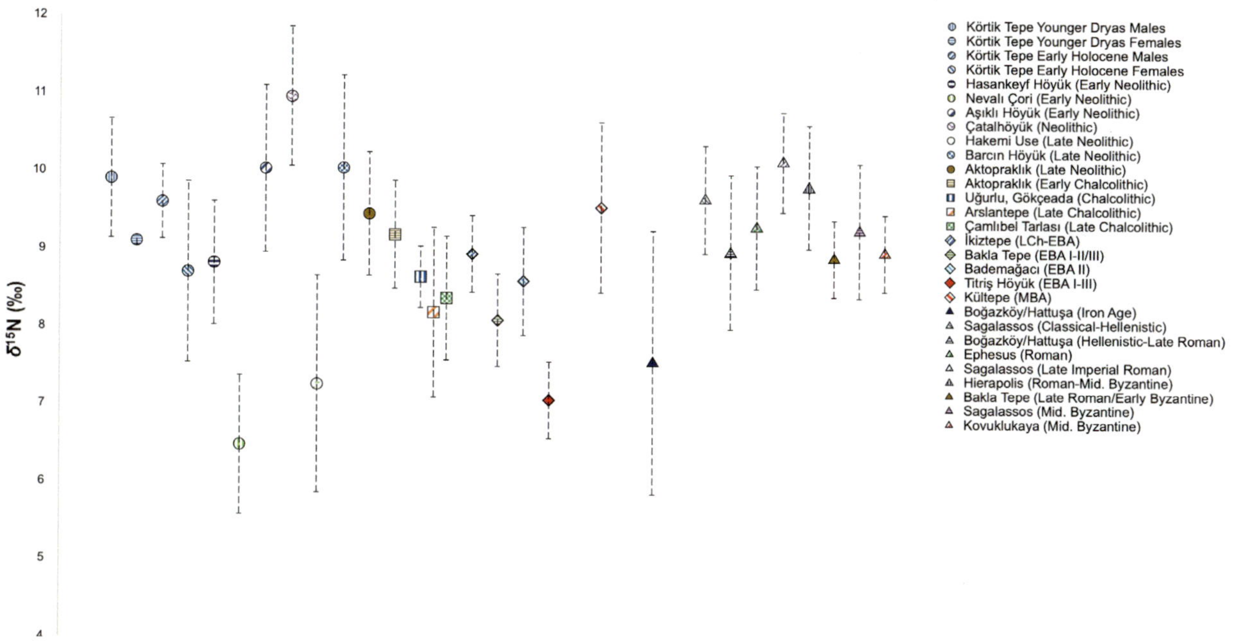

Fig. 10.5. Means and SD for $\delta^{15}N$ values from Anatolian populations through time. For references and sources of the data see Table 10.1 (chart by author).

Discussion

The stable isotope data indicates that dietary protein was from terrestrial sources, and that marine or aquatic food resources (despite the proximity of some sites near to the coast or rivers) never played an important role in Anatolian dietary habits. For the prehistoric periods this is not so surprising, as similar findings have been observed in EBA Anatolia (Irvine et al. 2019), prehistoric Greece (Papathanasiou, Richards 2015; Pokutta 2017), Starčevo (Neolithic) communities outside the Danube Gorges (Jovanović et al. 2019) and prehistoric populations from other regions (see discussions in Jovanović et al. 2019; Irvine, Erdal 2020). However, from written sources and some archaeozoological and archaeological research of the later, historical, periods we know that aquatic and marine resources supposedly played a more substantially visible role in the dietary habits of these periods. This was particularly due to improved methods of preservation to prevent spoiling, such as salting and drying and garum sauce (Feyer 1936; Apicius et al. 1958; Curtis 1991; Grant 2000; Schiefsky 2005; Łukaszewicz 2011; MacKinnon 2011; Aslan 2015; Teegen 2015; Sturtevant 2016). It may be that despite the written and material evidence, marine and aquatic resources did not actually play a big role in the daily dietary habits of historical period Anatolian populations. Another possibility may be related to the averaging effects of human bone collagen turnover, with stable isotope values reflecting the last ca 10–15 years of life, which could mask intermittent consumption of aquatic and/or marine resources. It may also be due to the fact that, whilst some of the historical period sites analysed in this study are within reach of the coast, none are directly on the coast or could be considered to be coastal settlements. Furthermore, aquatic resources can produce similar $\delta^{13}C$ and $\delta^{15}N$ signals to terrestrial ones (Triantaphyllou et al. 2006; Vika et al. 2009), and their consumption may be hidden for this reason. Research into human dietary habits using stable isotopes ($\delta^{13}C$ and $\delta^{15}N$) at Classical Period Thebes has suggested the consumption of freshwater fish, and in particular eels (Vika et al. 2009). This was due to some individuals at the site having enriched $\delta^{15}N$ values (>11‰), but without a corresponding $\delta^{13}C$ enrichment. Similar values are not observed in the Anatolian historical period sample populations, although some individuals do have $\delta^{15}N$ values around the 11‰ mark, so freshwater fish consumption is not impossible. Interestingly, some of the individuals from Early Neolithic Aşıklı Höyük and Neolithic Çatalhöyük have $\delta^{15}N$ values of ca 11‰ to 13‰, so perhaps some consumption of freshwater resources could be suggested, although the researchers did not suggest this as an option, and (for Çatalhöyük) instead indicated that the high $\delta^{15}N$ may be due to the consumption of cattle protein (Richards et al. 2003). Future research utilising isotopic or biogeochemical dietary markers or methods that can distinguish and identify these resources may help to shed more light on this issue.

	Neolithic	Chalcolithic	EBA	Classical-Byzantine
Neolithic	1	0.444	**<0.0001**	**<0.0001**
Chalcolithic	0.444	1	**0.0001**	**<0.0001**
EBA	**<0.0001**	**<0.0001**	1	**<0.0001**
Classical-Byzantine	**<0.0001**	**<0.0001**	**<0.0001**	1

Tab. 10.3. Pairwise comparison p-values for $\delta^{13}C$ between the chronologically grouped sample populations. Bold indicates statistically significant.

	Neolithic	Chalcolithic	EBA	Classical-Byzantine
Neolithic	1	0.051	**0.000**	0.133
Chalcolithic	0.051	1	0.258	**0.002**
EBA	**0.000**	0.258	1	**<0.0001**
Classical-Byzantine	0.133	**0.002**	**<0.0001**	1

Tab. 10.4. Pairwise comparison p-values for $\delta^{15}N$ between the chronologically grouped sample populations. Bold indicates statistically significant.

According to the statistical analyses the Classical-Byzantine sample population was the most different from the others for $\delta^{13}C$; this can also be clearly seen in Figure 10.3. This is not surprising though, as it is the only sample population group whose diet was likely not based on an entirely terrestrial C_3-based mixed diet. The deliberate introduction and cultivation of C_4 plants (primarily sorghum and millet) into the arable repertoire of Anatolia and surrounding regions is known from the LBA/Iron Age onwards, and was likely part of the dietary assemblage to varying degrees (Nesbitt, Summers 1988; Riehl 2009; Miller et al. 2016; Murphy 2016; Herrscher et al. 2018). In Figure 10.2 it is observable that there are several individuals with $\delta^{13}C$ values around, or more positive than the ca -18‰ value that has been accepted as a threshold to indicate increased C_4 input in human diet (Hedges 2003; Lee-Thorp 2008; Herrscher et al. 2018; Wang et al. 2019; Liu et al. 2020). Whilst it can also be seen in Figures 10.2 and 10.4 that some of the Neolithic and MBA Kültepe individuals also have values towards the -18‰ threshold, this is likely to be an indirect, non-deliberate consumption of C_4 plant material through the consumption of animals that had been grazing or foddered on wild C_4 grasses (Richards et al. 2003; Herrscher et al. 2018; Irvine et al. 2019; Rudov et al. 2020). Furthermore, the pairwise comparison test is based on the overall means of the chronologically grouped sample populations. From Figure 10.3, it can be observed that the means of the Neolithic and Chalcolithic sample populations, for $\delta^{13}C$ and $\delta^{15}N$, are nearly identical. However, in the Chalcolithic period there are lower ranges, and standard deviations around the means, and thereby dietary variation.

Due to the pairwise comparison test being based on the overall means of the chronologically grouped sample populations it is also important to consider the means of the populations within these chronological groupings. When these discrete means and ranges (Figs 10.4–10.5) are examined, they form what is effectively an 'hourglass' pattern; with greater variability and heterogeneity in the Neolithic and Classical-Byzantine periods, and a narrower range in the EBA, in particular. The pairwise comparison test for $\delta^{15}N$ actually indicates this with the EBA populations being significantly different from the Neolithic and Classical-Byzantine populations. This 'isotopic hourglass' is also reflected in the archaeobotanical and archaeozoological record. This can be very simply summarised (see Fig. 10.6) as a diversity of cultivated (and collected) crop species, and a mix of herding and hunting in the Neolithic (and Early Chalcolithic), followed by the standardised repertoire of the 'Mediterranean Quartet' consisting of cereals (wheat and barley), vines, olive and legumes (Sarpaki 1992), and sheep/goat, cattle and some pigs in the fourth-to-second millennia BC. In the post-Bronze Age periods the introduction and deliberate cultivation of additional key cereal crops (e.g. C_4 sorghum and millet), and the supposed (re)-introduction of an increased component of wild animal protein (specifically aquatic and marine resources) results in an increasingly diverse diet. It should be clarified here that in this instance, diversity alludes to not

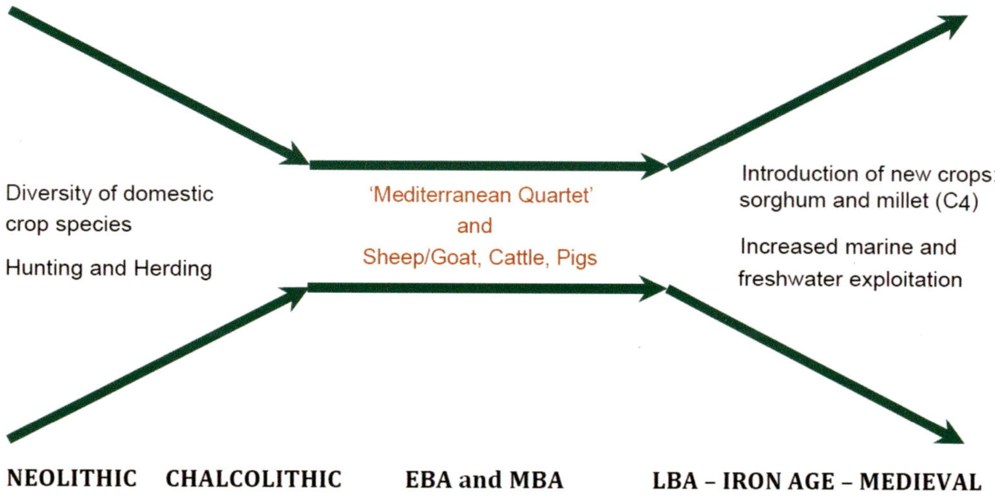

Fig. 10.6. Schematic representation of the 'hourglass' theory for changes in agriculture over time in Anatolia (figure by author).

only the variety of species of plants and/or animals, but, in particular, the potential for isotopic signal variety. Also, when reconstructing palaeodiets in this manner, it is only the 'key' crops/plant material and sources of animal protein that are considered. Basically, this refers to the foods that would likely have constituted the bulk of daily meals, rather than those that would have provided supplementary sources of nutrition (e.g. fruits, nuts, foraged shellfish or hunted birds/small game, etc.).

The observed reduction in the mean δ^{15}N values from the Neolithic to EBA (Fig. 10.5) may possibly be explained by the extensification of agriculture in the fourth millennium BC (Wilkinson 1994 and Paul Halstead's comments in reply; Bogaard et al. 2007; 2013; Styring et al. 2017). The extensification in agriculture would have resulted in a lower amount and intensity of crop manuring, and thereby a lowering of the δ^{15}N values at the base of the food chain (Bogaard et al. 2007; 2013; Styring et al. 2017). It is also possible that with an increased intensity in the exploitation of animals' ante-mortem products (particularly traction and wool), animals became viewed as valuable commodities, rather than simply as a source of food. This may have resulted in a general reduction of the amount of animal protein consumed on a daily basis, relative to other time periods (Atıcı 2014; Styring et al. 2017; Irvine et al. 2019).

The Classical-Byzantine period of Anatolia is actually quite distinct, and furthermore, its isotopic profile is not as heterogenous as initially expected with the (relatively) increased variety of available food sources (e.g. C_4 crops, and a wider range of domestic and wild animals supposedly exploited for food). The dietary habits and subsistence strategies of the post-Bronze Age periods of Anatolia, from an isotopic perspective, will be examined in more depth in an upcoming publication (Özdemir, Irvine, forthcoming).

Conclusion

This study utilises published stable isotope data (δ^{13}C and δ^{15}N) from human bone collagen to examine changing dietary habits over time in Anatolia. Thereby, it also provides preliminary implications about diachronic changes in subsistence practices and agricultural strategies. The examined data comes from a variety of sites, both temporally and geographically, within the Anatolian peninsula.

The key finding of this study suggests that there is an 'hourglass' effect with regard to agriculture and dietary habits in Anatolia. A wide range of human stable isotope values in the Anatolian Neolithic is concurrent with the varied strategies of early farming, and a diverse range of exploited and managed plant and animal food resources. A homogeneous range of human stable isotope data in the EBA is matched by a monoculture of the key domestic crops and animals, and a relatively standardised strategy of extensive agriculture. It is evident that the stable isotope data agrees with the archaeobotanical and archaeozoological evidence that there is a narrowing in the range of exploited food resources commencing in the Chalcolithic period and most evident by the third millennium BC. The EBA period is notable for being a period of intensified human-environment interactions and also of agricultural, architectural and cultural standardisation. There is also a clear pattern of agricultural extensification, as well as an intensification in animal management and exploitation. In the historical periods (Classical to Middle Byzantine), the deliberate introduction and addition of C_4 plants into the agricultural assemblage is noted through a change in the $\delta^{13}C$ values. Furthermore, the ability to preserve meat and marine resources allowed for the transportation of protein sources, in turn allowing for a potentially more varied input into the dietary habits, even if this is not immediately and clearly visible in the examination of the current stable isotope values. A general reduction in $\delta^{15}N$ values is observed from the Late Neolithic to EBA periods, followed by a slight increase in the following periods. This pattern is likely related to increasingly extensive agricultural strategies, and the intensification of the ante-mortem product exploitation of domestic animals.

Unfortunately, at the time of writing there are still relatively few stable isotope datasets available from Anatolia, and in particular from the Middle and Late Bronze Ages, and the Early Iron Age. And whilst this paper provides a preliminary overview, these gaps in the data record should be addressed to further examine changes in subsistence patterns and dietary habits through time in Anatolia.

Acknowledgements

I would like to sincerely thank Kameray Özdemir for his discussions and input, which have helped to develop and clarify some of the ideas discussed in this paper. I would also like to thank Michele Massa for producing the map, and for his remarks. My gratitude as well to Bike Yazıcıoğlu-Santamaria for granting me permission to use the data from her doctoral thesis, and for her comments on a draft of this paper. And, of course, I would like to extend my thanks to the anonymous reviewers who helped to improve the earlier versions of this paper.

Bibliography

Ambrose, S.H. 1993: 'Isotopic analysis of paleodiets: methodological and interpretive considerations' in M.K. Sandford (ed.), *Investigations of Ancient Human Tissue: Chemical Analyses in Anthropology*. Philadelphia, Gordon and Breach: 59–130

Ambrose, S.H., Norr, L. 1993: 'Experimental evidence for the relationship of the carbon isotope ratios of whole diet and dietary protein to those of bone collagen and carbonate' in J.B. Lambert, G. Grupe (eds), *Prehistoric Human Bone Archaeology at the Molecular Level*. Berlin, Springer Verlag: 1–38

Apicius, Flower, B., Alföldi-Rosenbaum, E. 1958: *The Roman Cookery Book: A Critical Translation of the Art of Cooking, for Use in the Study and the Kitchen*. London, Peter Nevill

Aslan, E. 2015: 'Kekova bölgesi'nde bulunan balik sosu işlikleri' *Cedrus* 3: 141–61

Atıcı, L. 2014: 'The secondary products revolution in light of textual evidence from Kültepe-Kanesh, central Turkey' in H. Greenfield (ed.), *Animal Secondary Products: Domestic Animal Exploitation in Prehistoric Europe, the Near East and the Far East*. Oxford, Oxbow Books: 233–52

Benz, M., Erdal, Y.S., Sahin, F., Özkaya, V., Alt, K. 2016: 'The equality of inequality – social differentiation among the hunter-fisher-gatherer community of Körtik Tepe, South-Eastern Turkey' in H. Meller, H.P. Hahn, R. Jung, and R. Risch (eds), *Arm und Reich – Zur Ressourcenverteilung in prähistorischen Gesellschaften (Rich and Poor – Competing for Resources in Prehistoric Societies): 8. Mitteldeutscher Archäologentag vom 22. bis 24. Oktober 2015 in Halle (Saale)*. Tagungen des Landesmuseums für Vorgeschichte Halle 14/1. Halle (Saale), Landesmuseum für Vorgeschichte: 147–64

Bogaard, A., Heaton, T.H.E., Poulton, P., Merbach, I. 2007: 'The impact of manuring on nitrogen isotope ratios in cereals: archaeological implications for reconstruction of diet and crop management practices' *Journal of Archaeological Science* 34.3: 335–43

Bogaard, A., Fraser, R., Heaton, T.H.E., Wallace, M., Vaiglova, P., Charles, M., Jones, G., Evershed, R.P., Styring, A.K., Andersen, N.H., Arbogast, R.-M., Bartosiewicz, L., Gardeisen, A., Kanstrup, M., Maier, U., Marinova, E., Ninov, L., Schäfer, M., Stephan, E. 2013: 'Crop manuring and intensive land management by Europe's first farmers' *Proceedings of the National Academy of Sciences* 110.21: 12589–94

Brami, M.N. 2019: 'Anatolia: from the origins of agriculture ... to the spread of Neolithic economies' in M. Brami, and B. Horejs. (eds), *The Central/Western Anatolian Farming Frontier. Proceedings of the Neolithic Workshop held at 10th ICAANE in Vienna, April 2016*. Vienna, Austrian Academy of Sciences Press: 17–43

Budd, C., Lillie, M., Alpaslan-Roodenberg, S., Karul, N., Pinhasi, R. 2013: 'Stable isotope analysis of Neolithic and Chalcolithic populations from Aktopraklık, northern Anatolia' *Journal of Archaeological Science* 40: 860–67

Budd, C., Karul, N., Alpaslan-Roodenberg, S., Galik, A., Schulting, R., Lillie, M. 2018: 'Diet uniformity at an early farming community in northwest Anatolia (Turkey): carbon and nitrogen isotope studies of bone collagen at Aktopraklık' *Archaeological and Anthropological Sciences* 10.8: 2123–35

Budd, C., Galik, A., Alpaslan-Roodenberg, S., Schulting, R., Lillie, M. 2020: 'Early farmers in northwest Turkey: first dietary isotopes study of human diet at Neolithic Barcın Höyük' *Journal of Archaeological Science: Reports* 31: 102288 https://doi.org/10.1016/j.jasrep.2020.102288

Curtis, R.I. 1991: *Garum and Salsamenta: Production and Commerce in Materia Medica*. Leiden, Brill

DeNiro, M.J. 1985: 'Postmortem preservation and alteration of in vivo bone collagen isotope ratios in relation to paleodietary reconstruction' *Nature* 317: 806–09

Faure, G., Mensing, T.M. 2005: *Isotopes Principles and Applications*, 3rd edition. Hoboken, John Wiley & Sons

Fernandes, R., Nadeau, M.-J., Grootes, P.M. 2012: 'Macronutrient-based model for dietary carbon routing in bone collagen and bioapatite' *Archaeological and Anthropological Sciences* 4: 291–301. https://doi.org/10.1007/s12520-012-0102-7

Feyer, M. 1936: 'Nouvelles inscriptions d'Akraiphia' *Bulletin de Correspondance Hellénique* 60: 27–36

Fuller, B.T., De Cupere, B., Marinova, E., Van Neer, W., Waelkens, M., Richards, M.P. 2012: 'Isotopic reconstruction of human diet and animal husbandry practices during the Classical-Hellenistic, Imperial, and Byzantine periods at Sagalassos, Turkey' *American Journal of Physical Anthropology* 149: 157–71

Gamarra, B., Howcroft, R., McCall, A., Dani, J., Hajdú, Z., Gyöngyvér Nagy, E., Szabó, L.D., Domboróczki, L., Pap, I., Raczky, P., Marcsik, A., Zoffmann, Z.K., Hajdu, T., Feeney, R.N.M., Pinhasi, R. 2018: '5000 years of dietary variations of prehistoric farmers in the Great Hungarian Plain' *PLoS ONE* 13.5: e0197214. https://doi.org/10.1371/journal.pone.0197214

Grant, M. 2000: *Galen on Food and Diet*. London, Routledge

Greenberg, R., Palumbi, G. 2015: 'Corridors and colonies: comparing fourth-third millennia BC interactions in southeast Anatolia and the Levant' in A.B. Knapp, P. van Dommelen (eds), *The Cambridge Prehistory of the Bronze and Iron Age Mediterranean*. Cambridge, Cambridge University Press: 111–38

Guiry, E.J., Szpak, P. 2021: 'Improved quality control criteria for stable carbon and nitrogen isotope measurements of ancient bone collagen' *Journal of Archaeological Science* 132: 105416. https://doi.org/10.1016/j.jas.2021.105416

Hedges, R.E.M. 2003: 'On bone collagen-apatite-carbonate isotopic relationships' *International Journal of Osteoarchaeology* 13: 66–79

Herrscher, E., André, G., Bodet, C., Chataigner, C., Decaix, A., Goude, G., Hamon, C., Le Mort, F., Lyonnet, B., Martin, L., Messager, E., Oberlin, C., Ollivier, V., Poulmarc'h, M., Sermet, C., Vila, E. 2018: 'The origins of millet cultivation in the Caucasus: archaeological and archaeometric approaches' *Préhistoires Méditerranéennes* 6. http://journals.openedition.org/pm/1367

Heyd, V., Aydıngün, Ş., Güldoğan, E. 2016: 'Kanlıgeçit – Selimpaşa – Mikhalich and the question of Anatolian colonies in Early Bronze Age Southeast Europe' in B.P.C. Molloy (ed.), *Of Odysseys and Oddities. Scales and Modes of Interaction between Prehistoric Aegean Societies and Their Neighbours*. Oxford, Oxbow Books: 169–202

Iacumin, P., Di Matteo, A., Macrì, A., Galli, E. 2020: 'Late Chalcolithic population in Arslantepe: stable isotope reconstruction of diet and subsistence strategies' *Origini* 44: 51–64

Irvine, B., Erdal, Y.S., Richards, M.P. 2019: 'Dietary habits in the Early Bronze Age (3rd millennium BC) of Anatolia: a multi-isotopic approach' *Journal of Archaeological Science: Reports* 24: 253–63

Irvine, B., Erdal, Y.S. 2020: 'Analysis of dietary habits in a prehistoric coastal population (İkiztepe, north Anatolia), using stable isotopes of carbon, nitrogen, and sulphur' *Journal of Archaeological Science: Reports* 29:102067

Irvine, B., Özdemir, K. 2020: 'Biogeochemical approaches to bioarchaeological research in Turkey: a review' *Journal of Eastern Mediterranean Archaeology and Heritage Studies* 8.2: 174–99

Işıklı, M., Altunkaynak, G. 2016: 'The role of eastern Anatolia in relationships between the southern Caucasus and the Aegean world: some metallurgical evidence from the highlands of Erzurum-Kars' in M. Kvachadze, N. Shanshashvili (eds), *Aegean World and South Caucasus: Cultural Relations in the Bronze Age. International Workshop September 23–25, 2016, Georgia*. Tbilisi, Mtsignobari: 112–21

Itahashi, Y., Miyake, Y., Maeda, O., Kondo, O., Hongo, H., Van Neer, W., Chikaraishi, Y., Ohkouchi, N., Yoneda, M. 2017: 'Preference for fish in a Neolithic hunter-gatherer community of the upper Tigris, elucidated by amino acid δ^{15}N analysis' *Journal of Archaeological Science* 82: 40–49

Itahashi, Y., Erdal, Y.S., Tekin, H., Omar, L., Miyake, Y., Chikaraishi, Y., Ohkouchi, N., Yoneda, M. 2019: 'Amino acid ^{15}N analysis reveals change in the importance of freshwater resources between the hunter-gatherer and farmer in the Neolithic upper Tigris' *American Journal of Physical Anthropology* 168.4: 676–86

Itahashi, Y., Stiner, M.C., Erdal, O.D., Duru, G., Erdal, Y.S., Miyake, Y., Güral, D., Yoneda, M., Özbaşaran, M. 2021: 'The impact of the transition from broad-spectrum hunting to sheep herding on meat consumption: multi-isotopic analyses of human bone collagen at Aşıklı Höyük, Turkey' *Journal of Archaeological Science* 136: 105505 https://doi.org/10.1016/j.jas.2021.105505

Jovanović, J., de Becdelièvre, C., Stefanović, S., Živaljević, I., Dimitrijević, V., Goude, G. 2019: 'Last hunters–first farmers: new insight into subsistence strategies in the Central Balkans through multi-isotopic analysis' *Archaeological and Anthropological Sciences* 11: 3279–98

Katzenberg, M.A. 2000: 'Stable isotope analysis: a tool for studying past diet, demography, and life history' in M.A. Katzenberg, S.R. Saunders (eds), *Biological Anthropology of the Human Skeleton*. New York, Wiley-Liss: 305–28

Larsen, C.S. 2018: 'Bioarchaeology in perspective: from classifications of the dead to conditions of the living' *American Journal of Physical Anthropology* 165: 865–78

Lee-Thorp J.A. 2008: 'On isotopes and old bones' *Archaeometry* 50: 925–50

Liu, R., Pollard, M., Schulting, R., Rawson, J., Liu, C. 2020: 'Synthesis of stable isotopic data for human bone collagen: a study of the broad dietary patterns across ancient China' *The Holocene* 31.2: 302–12 https://doi.org/10.1177/0959683620941168

Lösch, S., Grupe, G., Peters, J. 2006: 'Stable isotopes and dietary adaptations in humans and animals at pre-pottery Neolithic Nevalı Çori, southeast Anatolia' *American Journal of Physical Anthropology* 131: 181–93

Lösch, S., Moghaddam, N., Grossschmidt, K., Risser, D.U., Kanz, F. 2014: 'Stable isotope and trace element studies on gladiators and contemporary Romans from Ephesus (Turkey, 2nd and 3rd Ct. AD) – implications for differences in diet' *PLoS ONE* 9.10: e110489. https://doi.org/10.1371/journal.pone.0110489

Łukaszewicz, A. 2011: 'A fish from the sea' in P. Bieliński, K.M. Ciałowicz, W.A. Daszewski, M. Gawlikowski, W. Godlewski, K. Myśliwiec (eds) *Classica Orientalia: Essays Presented to Wiktor Andrzej Daszewski on His 75th Birthday*. Warsaw, Polish Centre of Mediterranean Archaeology University of Warsaw, Wydawnictwo DiG: 353–56

MacKinnon, M.R. 2011: 'Animal use at Hellenistic Pergamon: evidence from zooarchaeological analyses' *Archäologischer Anzeiger* 2011.2: 193–98

Massa, M., Palmisano, A. 2018: 'Change and continuity in long-distance exchange networks between western/central Anatolia, northern Levant and northern Mesopotamia, c. 3200–1600 BCE' *Journal of Anthropological Archaeology* 49: 65–87

Miller, N.F., Spengler, R.N., Frachetti, M. 2016: 'Millet cultivation across Eurasia: origins, spread, and the influence of seasonal climate' *The Holocene* 26.10: 1566–75

Murphy, C. 2016: 'Finding millet in the Roman world' *Archaeological and Anthropological Sciences* 8: 65–78

Nesbitt, M., Summers, G.D. 1988: 'Some recent discoveries of millet (Panicum miliaceum L. and Setaria (L.) P. Beauv.) at excavations in Turkey and Iran' *Anatolian Studies* 38: 85–97

Özdemir, K. 2018: 'Bizans Dönemi Kovuklukaya topluluğunun (Boyabat, Sinop) beslenme alışkanlıklarının yeniden yapılandırılması: sabit izotop oranı analizlerinin öncül sonuçları' *Hitit Üniversitesi Sosyal Bilimler Enstitüsü Dergisi* 11.2: 1155–76

Papathanasiou, A. 2015: 'Stable isotope analyses in Neolithic and Bronze Age Greece: an overview' in A. Papathanasiou, M.P. Richards, S.C. Fox (eds), *Archaeodiet in the Greek World: Dietary Reconstruction from Stable Isotope Analysis*. Princeton, The American School of Classical Studies at Athens: 25–55

Papathanasiou A., Richards, M.P. 2015: 'Summary: patterns in the carbon and nitrogen isotope data through time' in A. Papathanasiou, M.P. Richards, S.C. Fox (eds) *Archaeodiet in the Greek World: Dietary Reconstruction from Stable Isotope Analysis*. Princeton, The American School of Classical Studies at Athens: 195–203

Pickard, C., Schoop, U.-D., Dalton, A., Sayle, K.L., Channell, I., Calvey, K., Thomas, J.L., Bartosiewicz, L., Bonsall, C. 2016: 'Diet at Late Chalcolithic Çamlıbel Tarlası, North-Central Anatolia: an isotopic perspective' *Journal of Archaeological Science: Reports* 5: 296–306

Pickard, C., Caldeira, C., Harten, N., Üstündağ, H., Schoop, U.-D., Bartosiewicz, L., Schachner, A. 2017: 'Reconstructing Iron Age to Roman period diet from bioarchaeological remains: preliminary results from Boğazköy, north-central Anatolia' in A. Schachner (ed.), *Innovation Versus Beharrung Was Macht den Unterschied des Hethitischen Reichs im Anatolien des 2. Jahrtausends v. Chr.?: Internationaler Workshop Zu Ehren Von Jürgen Seeher Istanbul, 23–24 Mai 2014, BYZAS 23*. Istanbul, Ege Yayınları: 239–56

Pilaar Birch, S.E., Boz, B., Erdoğu, B. 2021: 'Isotope paleodietary analysis of humans from Chalcolithic Uğurlu, Gökçeada Island, Turkey' *Journal of Archaeological Science: Reports* 39: 103136. https://doi.org/10.1016/j.jasrep.2021.103136

Pokutta, D. 2017: 'Food, economy and social complexity in the Bronze Age world: a cross-cultural study' *Musaica archaeologica* 2: 22–40

Pollard, M., Batt, C., Stern, B., Young, S.M.M. 2007: *Analytical Chemistry in Archaeology*. Cambridge, Cambridge University Press

Quinn, C.P., Beck, J. 2016: 'Essential tensions: a framework for exploring inequality through mortuary archaeology and bioarchaeology' *Open Archaeology* 2: 18–41

Richards M.P., Hedges R.E.M. 1999: 'Stable isotope evidence for similarities in the types of marine foods used by Late Mesolithic humans at sites along the Atlantic coast of Europe' *Journal of Archaeological Science* 26: 717–22

Richards, M.P., Pearson, J.A., Molleson, T.I., Russell, N., Martin, L. 2003: 'Stable isotope evidence of diet at Neolithic Çatalhöyük, Turkey' *Journal of Archaeological Science* 30: 67–76

Riehl, S. 2009: 'Archaeobotanical evidence for the interrelationship of agricultural decision-making and climate change in the ancient Near East' *Quaternary International* 197: 93–114

Rudov, A., Mashkour, M., Djamali, M., Akhani, H. 2020: 'A review of C_4 plants in Southwest Asia: an ecological, geographical and taxonomical analysis of a region with a high diversity of C_4 eudicots' *Frontiers in Plant Science* 11:546518. https://doi.org/10.3389/fpls.2020.546518

Sarpaki, A. 1992: 'The palaeobotanical approach: the Mediterranean triad or is it a quartet?' in B. Wells (ed.), *Agriculture in Ancient Greece (Proceedings of the Seventh International Symposium at the Swedish Institute at Athens)*. Stockholm, Förlag: 61–76

Schiefsky, M.J. 2005: *Hippocrates: On Ancient Medicine (Studies in Ancient Medicine)*. Leiden, Brill

Schoeninger, M.J., DeNiro, M.J. 1984: 'Nitrogen and carbon isotopic composition of bone collagen from marine and terrestrial animals' *Geochimica et Cosmochimica Acta* 48: 625–39

Schwarcz, H.P., Schoeninger, M.J. 2011: 'Stable isotopes of carbon and nitrogen as tracers for paleo-diet reconstruction' in M. Baskaran (ed.) *Handbook of Environmental Isotope Geochemistry*, Vol. 1. Heidelberg, Springer: 725–42

Skourtanioti, E., Erdal, Y.S., Frangipane, M., Balossi Restelli, F., Aslıhan Yener, K., Pinnock, F., Matthiae, P., Özbal, R., Schoop, U.-D., Guliyev, F., Akhundov, T., Lyonnet, B., Hammer, E.L., Nugent, S.E., Burri, M., Neumann, G.U., Penske, S., Ingman, T., Akar, M., Shafiq, R., Palumbi, G., Eisenmann, S., D'Andrea, M., Rohrlach, A.B., Warinner, C., Jeong, C., Stockhammer, P.W., Haak, W., Krause, J. 2020: 'Genomic history of Neolithic to Bronze Age Anatolia, Northern Levant, and Southern Caucasus' *Cell* 181: 1158–75

Sturtevant, P.B. 2016: 'Eskiçağ'da Anadolu'nun Güney ve Batısında yer alan bölgelerde balık ve balıkçılık' *Adalya* 19: 75–93

Styring, A.K., Charles, M., Fantone, F., Hald, M.M., McMahon, A., Meadow, R.H., Nicholls, G.K., Patel, A.K., Pitre, M.C., Smith, A., Sołtysiak, A., Stein, G., Weber, J.A., Weiss, H., Bogaard, A. 2017: 'Isotope evidence for agricultural extensification reveals how the world's first cities were fed' *Nature Plants* 3: 17076. https://doi.org/10.1038/nplants.2017.76

Szpak, P., White, C.D., Longstaffe, F.J., Millaire, J.F., Sanchez, V.F.V. 2013: 'Carbon and nitrogen isotopic survey of northern Peruvian plants: baselines for paleodietary and paleoecological studies' *PLoS One* 8.1: 1–28

Şahoğlu, V. 2008: 'Crossing borders: the Izmir region as a bridge between the East and West during the Early Bronze Age' in C. Gillis, B. Sjöberg (eds), *Crossing Borders. Proceedings of the 7th, 8th, and 9th International Workshops, Athens 1997–1999*. Sävedalen, Paul Aström Förlag: 153–73

Teegen, W.R. 2015: 'Die anthropologisch-paläopathologischen Untersuchungen in Pergamon 2014' *Archäologischer Anzeiger* 2015.2: 158–63

Triantaphyllou S., Richards M.P., Touchais G., Philippa-Touchais A., Voutsaki S. 2006: 'Analyses of Middle Helladic skeleton material from Aspis, Argos: stable isotope analysis of human remains' *Bulletin de correspondance Hellénique* 130.2: 627–37

Vika, E., Aravantinos, V., Richards, M.P. 2009: 'Aristophanes and stable isotopes: a taste for freshwater fish in Classical Thebes (Greece)?' *Antiquity* 83: 1076–83 https://doi.org/10.1017/S0003598X00099361

Walker, P.L., DeNiro, M.J., 1986: 'Stable nitrogen and carbon isotope ratios in bone collagen as indices of prehistoric dietary dependence on marine and terrestrial resources in Southern California' *American Journal of Physical Anthropology* 71: 51–61

Wang, T., Wei, D., Chang, X., Yu, Z., Zhang, X., Wang, C., Hu, Y., Fuller, B.T. 2019: 'Tianshanbeilu and the isotopic millet road: reviewing the late Neolithic/Bronze Age radiation of human millet consumption from north China to Europe' *National Science Review* 6.5: 1024–39. https://doi.org/10.1093/nsr/nwx015

Wilkinson, T.J. 1994: 'The structure and dynamics of dry-farming states in Upper Mesopotamia [and comments and reply]' *Current Anthropology* 35.5: 483–520

Wong, M., Naumann, E., Jaouen, K., Richards, M.P. 2017: 'Isotopic investigations of human diet and mobility at the site of Hierapolis, Turkey' in J.R. Brandt, E. Hagelberg, G. Bjørnstad, and S. Ahrens (eds), *Life and Death in Asia Minor in Hellenistic, Roman and Byzantine Times: Studies in Archaeology and Bioarchaeology*. Oxford, Oxbow Books: 228–36

Yazıcıoğlu-Santamaria, G.B. 2015: The people of Kanesh: residential mobility, community life, and cultural pluralism in a Bronze Age city in Anatolia, Turkey. PhD thesis, University of Chicago, Chicago

Zimmermann, T. 2007: 'Anatolia as a bridge from north to south? Recent research in the Hatti heartland' *Anatolian Studies* 57: 65–75

11. Concluding remarks

Benjamin Irvine, Yılmaz Selim Erdal

One of the main aims of the 'Physical Anthropology in Anatolia' workshop held in November 2019 was to present the current state of biological anthropological studies in Türkiye and provide a snapshot of the studies in the field as primary examples. Turkish researchers who are specialists in their fields were invited to present papers containing general evaluations of methodologies and also case studies applying different techniques, taking into account the different fields of biological anthropology. Therefore, to some extent, the articles in this edited volume also present the current state of biological anthropological research in Türkiye.

This volume covers a wide range of topics, many of which complement and enhance each other. The papers by Serpil Eroğlu and Ayça Aydoğan et al., for example, explore different ways of examining biodistance and genetic relatedness and, together, provide illustrations of 'traditional' biological anthropological macroscopic examination of the skeletal remains and biomolecular methods (in this case aDNA). Both of these papers provide an important and lasting contribution to our understanding of population dynamics and movement in the past, and across large diachronic and geographical scales. This is not only within Anatolia, but how they were shaped, and indeed themselves shaped, genetic interactions further afield. Similarly, dietary habits and subsistence strategies are explored diachronically in Anatolian populations by M. Melis Koruyucu through the examination of dental caries, and Yılmaz Selim Erdal's paper explores the effects that the transition to agriculture had on the human skeleton. These physical and macroscopic analyses are complemented by Benjamin Irvine's paper through an examination of $\delta^{13}C$ and $\delta^{15}N$ values. These three papers, when read together, provide a holistic in-depth view of dietary habits and subsistence patterns, and of their diachronic changes in Anatolia. Whilst our understanding of dietary habits and subsistence practices in the past is prone to flux (related to the number and location of samples and the sizes of the resultant datasets), these papers present our current best understanding of them, as well as a greater comprehension of human-environment interactions, staple finance and agricultural economies and related population dynamics in ancient Anatolia.

The papers in this volume demonstrate how large-scale diachronic studies enhance our knowledge not only of changes in human behaviour and human-environment interactions through time, but also how these changes affected people at the individual, population, regional and pan-regional levels. They also help to demonstrate the added benefit and usefulness of applying more 'precise', specific quantifiable methodologies and techniques to research questions and studies to help develop our understanding of past human experiences. Almost all of these papers, through the fact that they represent the most current and up-to-date research and datasets of their respective subfields, provide meaningful and lasting contributions to biological anthropology in Türkiye. Many of them also represent state-of-the-art methodologies and techniques, which further emphasises their importance and lasting contribution. Therefore, it is anticipated that many of the contributions to this volume will become key reference papers for biological anthropological research, not just in Türkiye but in general.

However, what the papers also implicitly demonstrate is that, despite the advances in and adoption of these micro- and biomolecular techniques, there is still a place and importance in macroscopic examination of human skeletal remains when investigating anthropological aspects of the past. Much more than this, macroscopic analysis of human skeletal remains and palaeodemography and palaeopathology studies are crucial for better understanding and providing context to the data obtained from these quantifiable scientific methods. One of the key messages from this edited volume as a whole is that multifaceted and holistic approaches to exploring particular research agendas are not just important but essential. Whilst the individual papers in this volume may not necessarily employ a multifaceted or holistic approach, the combined reading of them does so. That is to say, 'there is more than one way to eat your yoghurt', and there are many ways to explore anthropological conditions, behaviours and experiences in past human populations. The more ways a skeletal individual or population is examined and investigated, the more complete our comprehension of them will be.

Biological anthropology in Türkiye has developed markedly since the days of primarily analysing skeletal and dental morphometrics and investigating race. This is particularly true since the latter part of the 20th century and definitely the 1990s, when studies have examined skeletal remains from Anatolia within broader bioarchaeological contexts and with more wide-ranging research questions. Research agendas have accelerated particularly in the last decade, with the introduction and application of new methodologies, including quantifiable scientific techniques, which has increased the ability to not only tackle existing and earlier research questions with more specificity and in more depth, but has enabled us also to tackle a greater variety of research questions, and to stimulate new ones. For example, whilst cranial and morphological measurements continue to be employed in Turkish biological anthropology, the research agendas surrounding them place the focus on examining biodistance and genetic and phenotypic mobility, drift and admixture. Researchers are no longer solely concerned with classifying and identifying these measurements, and osteological characteristics and differences between them, but how they are connected with population growth and movement, and intra- and inter-population dynamics. Examinations of dietary habits, subsistence practices, activities and pathologies are being examined from an osteobiographical and bioarchaeological perspective, investigating what they meant at the individual, the population, and the local and supra-regional scales. Furthermore, they are being supplemented and complemented by modern techniques such as biomolecular analyses – not just on human remains, but also on animal and plant residues – to obtain further and more detailed data that helps contribute to a greater overall and holistic understanding of human activities, life, inter-personal and inter-population interactions and dynamics, and human-biome and human-environment interactions in the past. Research is moving away from what may be considered the more 'abstract' to the more specific and definable.

Future directions

The importance of maintaining and encouraging the positive momentum of these last 10 years is critical. One extremely pertinent and valuable aspect of this is ensuring that the next generation of Turkish biological anthropologists are trained and equipped to tackle research in the modern world, as well as the integration and enhanced collaboration of both domestic and foreign researchers and specialists on projects. Students entering biological anthropology programmes need to be taught and trained in a variety of different methodologies, theories and techniques that build upon the conventional methods of anatomical, palaeodemographic and palaeopathological analyses following external macroscopic examination of human skeletal and dental remains. Furthermore, it is necessary for students to have access not only to different specialists (even within biological anthropology, e.g. specialists in dental remains, entheseal changes, pathological conditions and lesions, etc.) and their expertise, but also equipment, laboratories, literature and skeletal materials. This will enable them to enhance their ability to learn and gain knowledge and skill sets, as well as develop their critical analytical thinking through the development of their own research questions and projects so that they can ultimately forge their own career paths and continue the cycle of knowledge transfer. Another important aspect of this is allowing and providing for students to gain access to discussion and discourse, to hone their skills, and exchange ideas and techniques and formulate collaborations. One key way to achieve this is to create opportunities where they may present data and research, through conferences, symposia and workshops, as well as through publication.

Despite the very positive and important developments in the field of biological anthropology as a whole in Türkiye, there are still some limitations that need to be addressed. A prime example is the way in which data from skeletal individuals and populations are presented (both orally and in writing). Too often only demographic statistics (sex, age-at-death, binary presence or absence, and percentages of pathological and traumatic injury conditions and lesions) are given. Whilst these kinds of 'basic statistics' are essential for providing a basic understanding and context of the skeletal individual(s) and/or populations being examined, they are only a starting point, laying foundations to be applied to research questions. Critical analysis around what these demographic observations and statistics mean in terms of osteobiographies, and past human behaviours and human-environment and inter-population interactions, is too often absent, or at least underdeveloped. Hopefully, this volume and the papers in it provide a stepping stone demonstrating that whilst we must pay attention to, and not forget, the fundamentals of the discipline, the potential and utility in applying these fundamentals to further developed and evolving research questions is of utmost importance and benefit.

One final comment to make is that, as the discipline of biological anthropology becomes further integrated into archaeology and the archaeological, environmental, biological and geochemical sciences more broadly – and with an ever-greater overlap between them – the importance and viability of collaboration and multifaceted analyses needs to be given higher status. Whilst, as previously mentioned, this applies to the integration and collaboration between biological anthropologists and specialists of other disciplines and methods, both domestic and foreign, it also applies at a very basic level to archaeological excavations, and to the removal and treatment of human remains. As a community, we need to greater emphasise the importance of having a biological anthropologist on site, or at the very least of having a biological anthropologist train archaeologists and excavators in basic methodologies for the excavation, removal, recording and storage of human remains. This will benefit all involved, from the site director who wants to know more about the population(s) of the settlement they are excavating, to the biological anthropologists who will have to examine and analyse the skeletal remains (often at a later date in a different place), as well as any specialists who will later apply scientific methods to the skeletal remains (e.g. stable isotopes and/or aDNA).

Biological anthropology has always played an important role in analysing the people of ancient Anatolia, and this importance and relevance is only increasing as time goes on. Türkiye is one of the richest countries in the world for archaeological sites and remains, and biological anthropologists have been, are and will always be at the forefront of understanding the people that populated these ancient lands, as well as their behaviours, habits and interactions.

Thank you and acknowledgements

There are many people and institutions that we owe thanks to for the coming to fruition of this edited volume. Firstly, thank you to all of the participants in the original workshop held at the BIAA in November 2019, which started this process. Thank you for presenting your research, and then for agreeing to be part of the publication process by contributing a chapter to this edited volume. Thank you as well for your patience and understanding as we've worked to bring this together. We would like to also thank the BIAA for their hospitality, and for enabling the very useful and stimulating workshop that has ultimately led to the formulation of this book. At the BIAA we would like to thank Gülgün Girdivan and Başak Bodur in particular, for helping with administrative and organisational details. The BIAA Archaeological Monograph Series' publication and editorial team, and especially Tamar Hodos, Catherine Draycott, Abby Robinson and Janine Su must also be thanked for facilitating the development of this volume, and for offering us welcome advice and critique at crucial times. And finally, thank you to the reviewers who put in the time and effort to review each chapter and the book as a whole, and to improve its final quality.

Index

T

U